Collins

Cambridge Lower Secondary

Science

**PROGRESS BOOK 9:
STUDENT'S BOOK**

Series editor: David Martindill
Authors: Aidan Gill, Emma Poole and Heidi Foxford

William Collins' dream of knowledge for all began with the publication of his first book in 1819.

A self-educated mill worker, he not only enriched millions of lives, but also founded a flourishing publishing house. Today, staying true to this spirit, Collins books are packed with inspiration, innovation and practical expertise. They place you at the centre of a world of possibility and give you exactly what you need to explore it.

Collins. Freedom to teach.

Published by Collins

An imprint of HarperCollins*Publishers*
The News Building, 1 London Bridge Street, London, SE1 9GF, UK

HarperCollins*Publishers*
Macken House, 39/40 Mayor Street Upper, Dublin 1, D01 C9W8, Ireland

Browse the complete Collins catalogue at
collins.co.uk

© HarperCollins*Publishers* Limited 2024

10 9 8 7 6 5 4 3 2 1

ISBN 978-0-00-867934-7

All rights reserved. No part of this publication may be reproduced, stored in a retrieval system, or transmitted in any form by any means, electronic, mechanical, photocopying, recording or otherwise, without the prior written permission of the Publisher or a licence permitting restricted copying in the United Kingdom issued by the Copyright Licensing Agency Ltd, 5th Floor, Shackleton House, 4 Battle Bridge Lane, London SE1 2HX.

British Library Cataloguing-in-Publication Data

A catalogue record for this publication is available from the British Library.

The questions, accompanying marks and mark schemes included in this resource have been written by the author and are for guidance only. They do not replicate examination papers and the questions in this resource will not appear in your exams. In examinations the way marks are awarded may be different. Any references to assessment and/or assessment preparation are the author's interpretation of the syllabus requirements.

This text has not been through the endorsement process for the Cambridge Pathway. Any references or materials related to answers, grades, papers or examinations are based on the opinion of the author. The Cambridge International Education syllabus or curriculum framework associated assessment guidance material and specimen papers should always be referred to for definitive guidance.

Series Editor: David Martindill
Authors: Aidan Gill, Emma Poole and Heidi Foxford
Publisher: Elaine Higgleton
Product manager: Catherine Martin
Product developer: Roisin Leahy
External Project Manager: Just Content Ltd
Development editor: Rebecca Ramsden
Copyeditor: Nick Hamar
Proofreader: Just Content Ltd
Cover designer: Gordon MacGilp
Cover illustrator: Ann Paganuzzi
Typesetter: PDQ Digital Media Solutions Ltd
Production controller: Sarah Hovell and Lyndsey Rogers
Printed and bound by Martins the Printers

We are grateful to the following teachers for providing feedback on the resources as they were developed: Dr. Rahul Sharma at IRA Global School, Mumbai, Mr Frank Akrofi and Mr Samuel Yeboah, Dániel Szücs at International School of Budapest, Ms Shalini Reddy at Manthan International School and Ms Sejal Vasrkar at SVKM JV Parekh International.

This book contains FSC™ certified paper and other controlled sources to ensure responsible forest management.

For more information visit: www.harpercollins.co.uk/green

Contents

Introduction	v
Photosynthesis and plant growth	1
Self-assessment and reflective learning page	5
Structure and function	7
Self-assessment and reflective learning page	10
Natural selection and inheritance	12
Self-assessment and reflective learning page	16
Periodic Table and salts	18
Self-assessment and reflective learning page	23
Chemical bonding	25
Self-assessment and reflective learning page	28
Reactions and rates	30
Self-assessment and reflective learning page	33
Energy and density	35
Self-assessment and reflective learning page	40
Electricity and sound	42
Self-assessment and reflective learning page	47
Earth and space	49
Self-assessment and reflective learning page	54
End of Year Test 1	56
End of Year Test 2	69
Periodic Table	82
Glossary	83

Introduction

This *Stage 9 Progress Student's Book* (and the *Stage 9 Progress Teacher Pack*) can be used to support the *Collins Cambridge Stage 9 Lower Secondary Science course* or to supplement your own resources. The *Progress Student's Book* contains

- nine End of Unit Tests offering practice questions to assess understanding of the Lower Secondary Science course
- two summative End of Year Tests
- Self-assessment sheets for each of the End of Unit Tests.

How to use the Progress resources

This downloadable, editable and photocopiable Student's Book contains a range of End of Unit Tests that are designed to be valuable and flexible formative and summative assessment resources. They can be used to identify the areas you are most confident in and to pinpoint how your teacher can support you to gain confidence in other areas.

The nine End of Unit Tests can be used as class tests or can be taken home to complete in your own time. They can be set at the end of a unit of teaching or can be combined to create a longer end of term test if appropriate.

Some of the questions in each End of Unit Test are written to address the Cambridge Thinking and Working Scientifically Learning Objectives:

- Models and representations
- Scientific enquiry: purpose and planning
- Carrying out scientific enquiry
- Scientific enquiry: analysis, evaluation and conclusions.

Each End of Unit Test is designed to be marked out of 20. Your teacher may set you a time limit of 20 minutes to complete these.

The End of Year Tests assess objectives taught across the whole year. The style of the End of Year Tests is otherwise the same as the End of Unit Tests, with a mixture of question styles and question difficulties, as well as the inclusion of some Thinking and Working Scientifically questions. Questions are also set in the context of practicals where appropriate, ensuring that you have experience of answering questions on investigative work. These tests could be used for summative purposes as end of year examinations or as practice to support you ahead of your examinations. The End of Unit Tests are separated out into the different science subjects, but the End of Year Tests cover a combination of the different science subjects.

The Self-assessment sheets give you the opportunity to reflect on your understanding. These contain a list of statements to judge your understanding of the course content where you are able to rank your understanding of the statement between 'I don't know', 'I need more practice' and 'I understand.' This will provide you with a relatively quick way to assess your overall confidence with the content and will also allow your teacher to plan how best to support you moving forwards. There is also space for you to produce a written reflection to answer these questions: 'What went well in this topic?', 'What could you do better next time?' and 'What parts of the course could your teacher go through in a revision lesson which would support your improvement?'. There is also a space for teachers to make a comment about your understanding of the content.

Teachers can use the results of the End of Unit Tests and the students' Self-assessment sheets to help them in future lesson planning. For example, if many students struggled with work linked to the internal structure of the Earth, a teacher may wish to bear this in mind when planning their teaching of a related topic, such as seismic waves – teachers could, for instance, include starter activities recapping the earlier work.

Key features: End of Unit and End of Year Tests

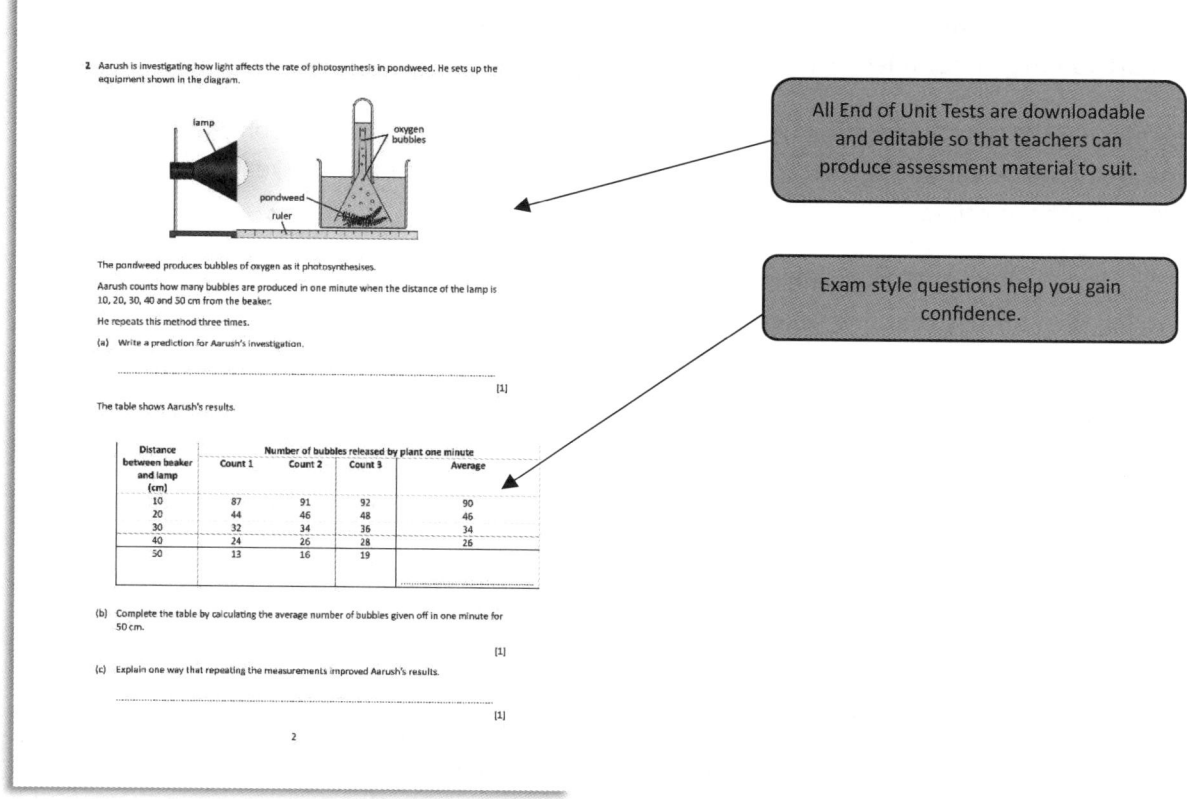

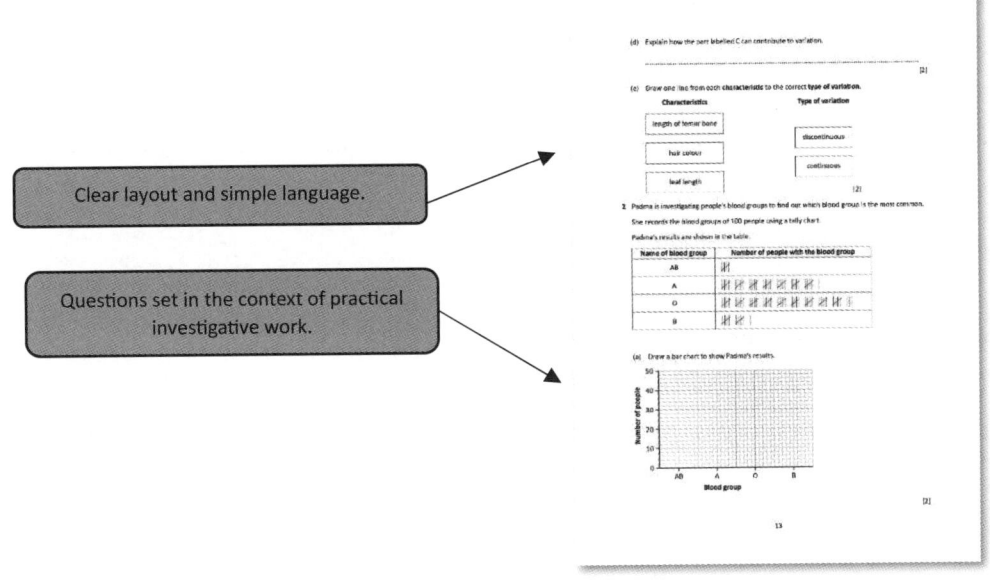

Clear layout and simple language.

Questions set in the context of practical investigative work.

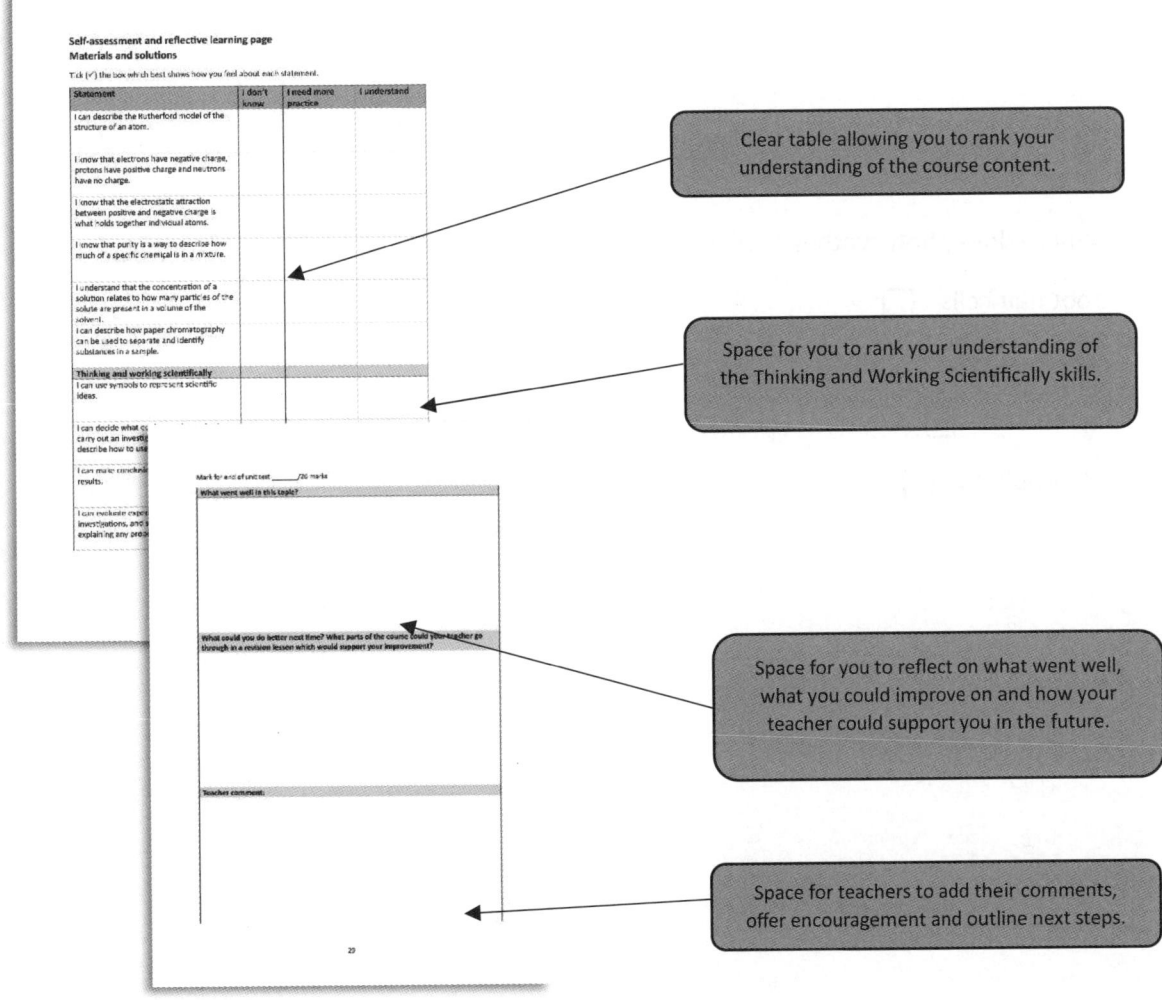

Clear table allowing you to rank your understanding of the course content.

Space for you to rank your understanding of the Thinking and Working Scientifically skills.

Space for you to reflect on what went well, what you could improve on and how your teacher could support you in the future.

Space for teachers to add their comments, offer encouragement and outline next steps.

End of Unit Test: Photosynthesis and plant growth
Total = 20 marks

Name: .... Class: ....
Date: ....

1 (a) Complete the sentences about photosynthesis.

Photosynthesis uses energy from the ....

Photosynthesis is a process that makes .... for the plant.

[2]

(b) Complete the word equation for photosynthesis.

carbon dioxide + ... → .... + oxygen

[1]

(c) Where does photosynthesis take place? Tick (✓) the correct box.

root hair cells ☐

xylem ☐

cytoplasm ☐

chloroplasts ☐

[1]

2 Aarush is investigating how light affects the rate of photosynthesis in pondweed. He sets up the equipment shown in the diagram.

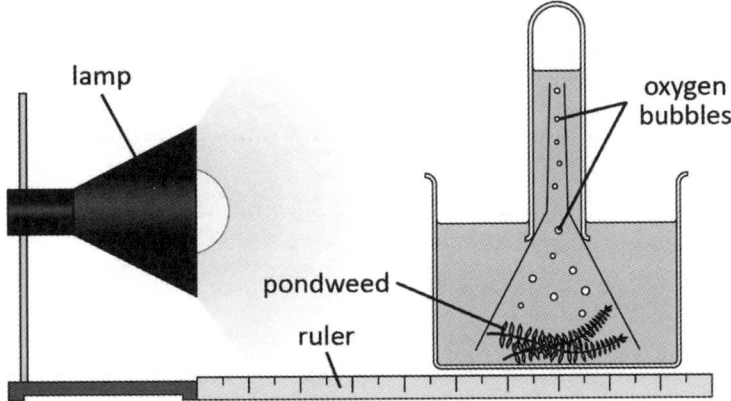

The pondweed produces bubbles of oxygen as it photosynthesises.

Aarush counts how many bubbles are produced in one minute when the distance of the lamp is 10, 20, 30, 40 and 50 cm from the beaker.

He repeats this method three times.

(a) Write a prediction for Aarush's investigation.

....
[1]

The table shows Aarush's results.

Distance between beaker and lamp (cm)	Number of bubbles released by plant one minute			
	Count 1	Count 2	Count 3	Average
10	87	91	92	90
20	44	46	48	46
30	32	34	36	34
40	24	26	28	26
50	13	16	19	....

(b) Complete the table by calculating the average number of bubbles given off in one minute for 50 cm.

[1]

(c) Explain one way that repeating the measurements improved Aarush's results.

....
[1]

Aarush's friend Ryan says it would be better to use a gas syringe to measure the volume of oxygen produced.

(d) Explain why measuring the volume of oxygen rather than counting the number of bubbles would improve the results.

....

....

[2]

3 Padma grows tomato plants.

She adds a plant fertiliser containing nitrates and magnesium to her tomato plants.

(a) Explain why tomato plants need nitrates and magnesium.

nitrates: ....

....

magnesium: ....

....

[2]

(b) Describe the pathway taken by a nitrate ion from the soil to the flowering parts of the plant.

....

....

....

....

[2]

4 The diagram below shows a cross-section of a leaf.

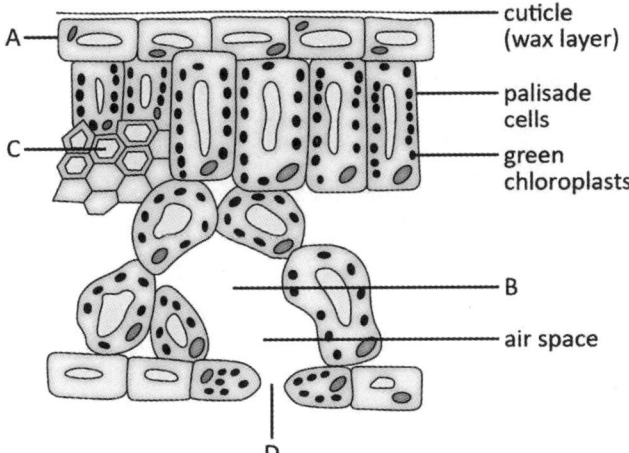

Small pores called stomata (singular stoma) in the leaves allow gases to move in and out of the leaf.

(a) Give the letter on the diagram that shows a stoma.

....
[1]

(b) Name the gas that moves into the stoma so it can be used for photosynthesis.

....
[1]

(c) The palisade cells have high numbers of chloroplasts. Name the pigment found in chloroplasts.

....
[1]

During transpiration, water vapour particles move out of the leaf by diffusion.

(d) Describe the process of the diffusion of water vapour from inside the leaf to the surrounding air.

....

....
[2]

(e) Plants that have fewer stomata are more able to survive long periods of hot dry weather.

Explain how having fewer stomata help can help a plant survive during hot and dry weather.

....

....
[2]

Self-assessment and reflective learning page
Photosynthesis and plant growth

Tick (✓) the box which best shows how you feel about each statement.

Statement	I don't know	I need more practice	I understand
I know that plants make carbohydrates using carbon dioxide, water and energy from the Sun, in a chemical reaction called photosynthesis and that photosynthesis happens in chloroplasts, containing chlorophyll, found inside some plant cells.			
I can write and use the word equation for photosynthesis.			
I can describe and explain how changing the light intensity alters the rate of photosynthesis.			
I know that plants require minerals to maintain healthy growth and life processes (nitrates to make proteins and magnesium to make chlorophyll).			
I can describe the route that water and mineral salts take from the roots to the leaves in flowering plants, including absorption in root hair cells, transport through xylem and transpiration from the surface of leaves.			
Thinking and working scientifically			
I can make a prediction using my knowledge of science.			
I can decide when to increase the range of observations and measurements, and increase the extent of repetition, to give sufficiently reliable data.			
I can describe trends and patterns in results.			

I understand that taking precise and accurate results is important to get results that are valid and reliable.	6		

Mark for end of unit test _____/20 marks

What went well in this topic?

What could you do better next time? What parts of the course could your teacher go through in a revision lesson which would support your improvement?

Teacher comment:

# End of Unit Test: Structure and function
Total = 20 marks

Name: .... Class: ....

Date: ....

1 The diagram shows a model of a part of a human organ system.

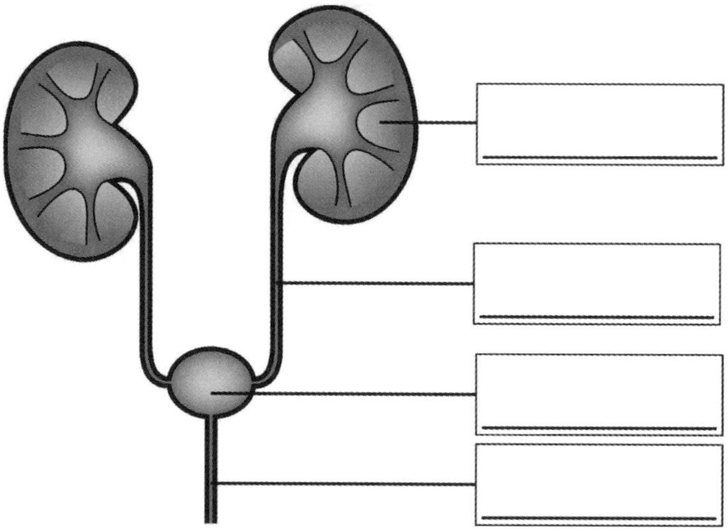

(a) Label the diagram using words from the list.

      bladder   **kidney**   **liver**   **lung**

      renal artery   **ureter**   **urethra**

[4]

(b) Name the organ system responsible for filtering the blood.

....

[1]

(c) A person can live a healthy and happy life with just one of their kidneys working.

However, if both the kidneys are damaged by a disease or accident the person will require extensive treatment to keep them alive.

Which statement best describes why a person with diseased or damaged kidneys is at risk of dying? Tick (✓) **one** box.

Because their kidneys would remove too much water from the body. ☐

Because the kidneys are responsible for controlling the amount of sugar in the blood. ☐

Because the kidneys filter the blood and remove waste substances such as urea. ☐

Because the kidneys are involved in the removal of carbon dioxide from the body. ☐

[1]

2 (a) Complete the sentences about sperm and egg cells.

Sperm and egg cells are known as ....

These specialised cells have only half the number of ....as 'normal' cells.

The sperm cell from a male fuses with an egg cell from a female to form a ball of cells that develops into a foetus. This process is known as ....

[3]

(b) Match each family member with their sex chromosomes.

Family member	Sex chromosomes
father	
daughter	XY
mother	XX
son	

[2]

(c) Most body cells have 46 chromosomes. Write down how many chromosomes are in egg cells.

....

[1]

(d) A man and a woman have two daughters. The woman is pregnant with a third child.

What is the probability (chance) that this baby will also be a girl?

Write your answer as a percentage.

....

[1]

3 A university investigated pregnant mothers who smoke. Scientists at the university monitored a group of 500 pregnant women.

- 250 of the pregnant women were non-smokers.
- 250 of the pregnant women regularly smoked during pregnancy.

The table shows data for the babies born to the two groups of pregnant mothers.

Effect on birth	Non-smokers	Smokers
average birth weight (grams)	3410	2968
premature births (% born before due date)	9	21

(a) Use the data in the table to describe the effect of smoking on the birth weight and the percentage of premature births.

....

....
[2]

(b) Calculate the percentage of non-smokers in the study who had a premature birth.

Show your working.

..................................%
[2]

(c) Name **one** variable, other than smoking, that could have affected birth weight of both groups of pregnant mothers.

....
[1]

(d) Explain why it is difficult to make a valid conclusion about the link between the effect of smoking on birth weight and the percentage of babies born prematurely.

....

....
[2]

Self-assessment and reflective learning page
Structure and function

Tick (✓) the box which best shows how you feel about each statement.

Statement	I don't know	I need more practice	I understand
I can describe the structure and function of the human excretory system and explain how the renal systems plays an important role in excretion by removing urea and waste substances from the body.			
I know that females produce gametes called egg cells and males produce gametes called sperm cells and I can describe the fusion of gametes to produce a fertilised egg with a new combination of DNA.			
I can describe the process of fertilisation and the fusion of the male and female gametes to form a ball of cells which becomes an embryo and I can describe the inheritance of sex in humans in terms of XX and XY chromosomes.			
I can discuss how foetal development is affected by the health of the mother, including the effect of diet, smoking and drugs.			
Thinking and working scientifically			
I can use symbols to represent scientific ideas.			
I can evaluate the strength of the evidence collected and how it supports, or refutes, the prediction.			
I can describe trends and patterns in results.			
I can make conclusions by interpreting results and explain the limitations of the conclusions.			

Mark for end of unit test _____ /20 marks

What went well in this topic?

What could you do better next time? What parts of the course could your teacher go through in a revision lesson which would support your improvement?

Teacher comment:

End of Unit Test: Natural selection and inheritance
Total = 20 marks

Name: .... Class: ....
Date: ....

1 (a) Complete the sentences using the words below.

chromosomes   cytoplasm   DNA   enzymes   genes   nucleus

Genetic material is found within the .... of the cell.

The genetic material is organised into structures called ....

These structures are made from a chemical called ....

[3]

The diagram shows a model of the genetic material found inside the nucleus.

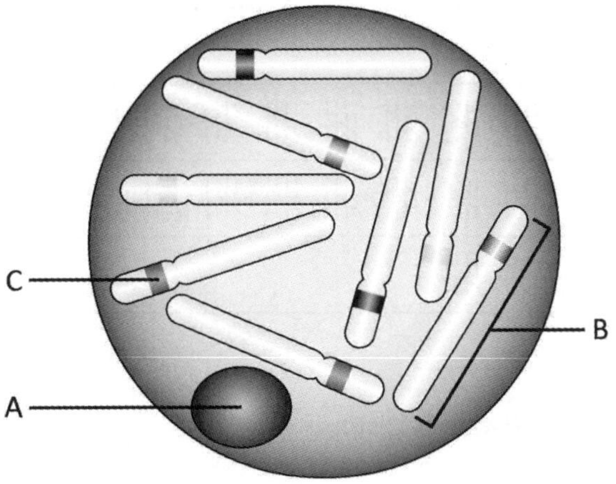

(b) In this model which letter represents:

   (i)   a chromosome? ....

   (ii)   a gene? ....

[2]

(c) Describe what is meant by the term variation.

....

....

[1]

(d) Explain how the part labelled C can contribute to variation.

....

[2]

(e) Draw one line from each **characteristic** to the correct **type of variation**.

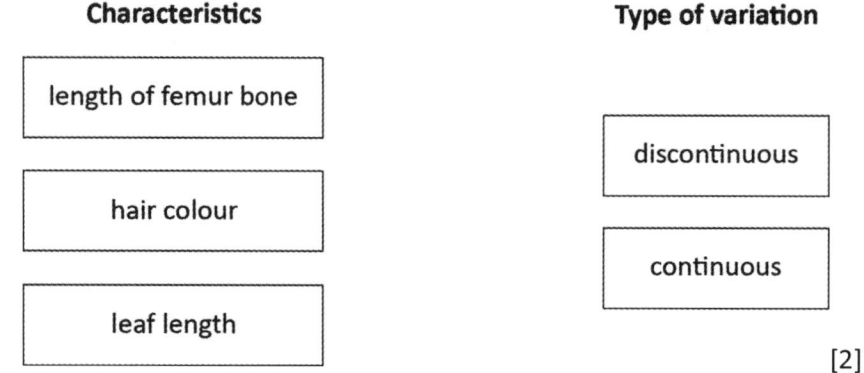

[2]

2 Padma is investigating people's blood groups to find out which blood group is the most common.

She records the blood groups of 100 people using a tally chart.

Padma's results are shown in the table.

Name of blood group	Number of people with the blood group
AB	ⲉⲕⲧ
A	ⲉⲕⲧ ⲉⲕⲧ ⲉⲕⲧ ⲉⲕⲧ ⲉⲕⲧ ⲉⲕⲧ ⲉⲕⲧ I
O	ⲉⲕⲧ ⲉⲕⲧ ⲉⲕⲧ ⲉⲕⲧ ⲉⲕⲧ ⲉⲕⲧ ⲉⲕⲧ ⲉⲕⲧ ⲉⲕⲧ III
B	ⲉⲕⲧ ⲉⲕⲧ I

(a) Draw a bar chart to show Padma's results.

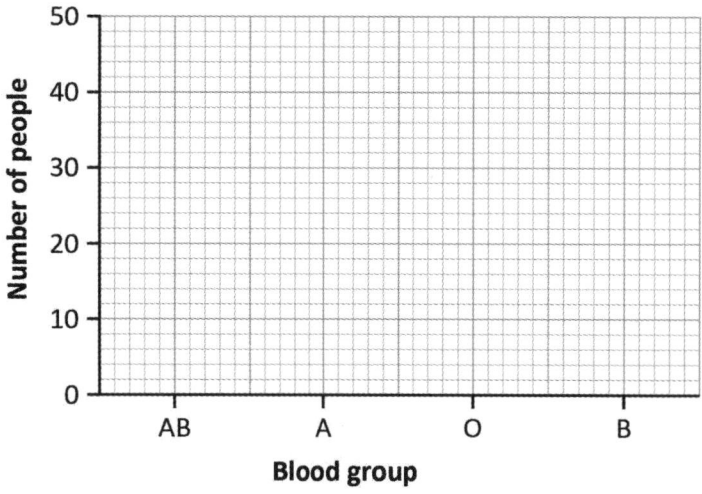

[2]

13

(b) Name the **two** most common blood groups in the group of people Padma investigated.

....
[1]

(c) Explain what causes different people to have different blood groups.

....
[1]

3 The golden toad has only ever been found in the Monteverde cloud forest in Costa Rica.

Between 1987 and 1991 scientists recorded the number of golden toads they observed.

Their results are shown in the table.

Year	Number of golden toads observed
1987	1500
1988	10
1989	1
1990	0
1991	0

(a) Calculate the percentage decrease in the population of the golden toad between 1987 and 1988.

....
[1]

(b) Identify the evidence that supports the claim that the golden toads are extinct.

....
[1]

(c) What is a possible reason the golden toads became extinct? Tick (✓) one answer.

an increase in resources ☐

less competition for mates ☐

a change in habitat due to climate change ☐

less competition for habitats ☐

[1]

4 Over many generations, peacocks (male) have evolved to have longer and more colourful tail feathers. It is thought that long and colourful tail feathers are more attractive to peahens (female).

Explain how populations of peacocks have evolved to have longer and more colourful feathers.

....

....

....

....

....

....

[3]

Self-assessment and reflective learning page
Natural selection and inheritance

Tick (✓) the box which best shows how you feel about each statement.

Statement	I don't know	I need more practice	I understand
I know that chromosomes contain genes, made of DNA, and that genes contribute to the determination of an organism's characteristics.			
I can identify the difference between a a chromosome, a gene and DNA in model diagrams of genetic material.			
I can describe variation within a species and relate this to genetic differences between individuals.			
I know that genes, along with the environment, determine variation in a species.			
I can describe continuous variation is a characteristic that changes gradually over a range of values such as height and discontinuous variation as a characteristic that has a distinct range of options or categories, such as eye colour.			
I can describe the scientific theory of natural selection and how it relates to genetic changes over time.			
I can describe what could happen to the population of a species, including extinction, when there is an environmental change.			

Thinking and working scientifically			
I know how to record and interpret measurements and data in an appropriate form.	17		
I can evaluate the strength of the evidence collected and how it supports, or refutes, the prediction.			
I can describe trends and patterns in results.			
I can present and interpret results.			

Mark for end of unit test _____/20 marks

What went well in this topic?

What could you do better next time? What parts of the course could your teacher go through in a revision lesson which would support your improvement?

Teacher comment:

# End of Unit Test: Periodic Table and salts
Total = 20 marks

Name: .... Class: ....

Date: ....

1 Keith wants to produce some crystals of copper sulfate.

   He adds some copper carbonate powder to a beaker containing 25.0 cm³ of warm sulfuric acid. He produces the salt copper sulfate and two other products.

   (a) Write a word equation for this reaction.

   ....

   [2]

   (b) (i) Name the piece of apparatus Keith should use to measure out exactly 25.0 cm³ of sulfuric acid.

   ....

   [1]

   (ii) Describe how Keith could produce a sample of pure dry copper sulfate from powdered copper carbonate and a solution of sulfuric acid.

   ....

   ....

   ....

   ....

   [4]

   (iii) Keith then repeats the experiment using 30 cm³ of sulfuric acid instead of 25 cm³ of sulfuric acid. Calculate the percentage increase in the volume of sulfuric acid used.

   ....%

   [2]

2 The Periodic Table shows all the known elements.

(a) Describe how the elements are arranged in the modern Periodic Table.

....

....
[1]

(b) Use your Periodic Table to write down the chemical symbol for magnesium.

....
[1]

(c) Identify the **group** of the Periodic Table that magnesium belongs to.

....
[1]

(d) Explain why magnesium belongs to **Period 3** of the Periodic Table.

....

....
[1]

The table below shows the observations of some Group 1 metals with water.

Group 1 metal	Observations when the metal is added to water
lithium	The metal floats on the water. There are a few bubbles.
sodium	The metal floats on water. It moves around and forms a ball. There are many bubbles.
potassium	The metal floats on water. There are lots of bubbles. The metal burns with a lilac flame.

(e) Use the information above and your Periodic Table to describe the trend in the reactions of Group 1 metals with water.

....

....
[1]

(f) Predict how the Group 1 metal **rubidium** would react with water.

....

....
[1]

3 The table below shows the density of elements in Group 1 of the Periodic Table. A student plots this data on a graph.

Group 1 metal	Density (g/cm³)
lithium	0.53
sodium	0.97
potassium	0.86
rubidium	1.53
caesium	1.87

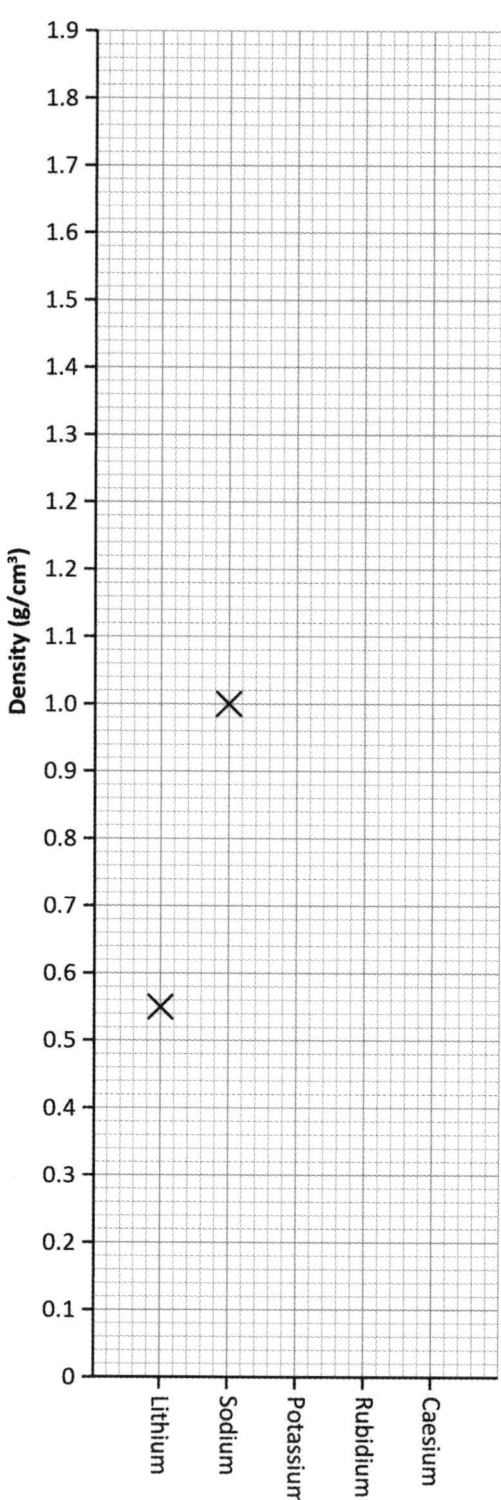

(a) Complete the graph:

   (i) Label the **x-axis**.

   [1]

   (ii) Plot the densities of the missing elements.

   [1]

(b) Identify the element with anomalous data and explain your answer.

....

....
[1]

(c) A lithium atom has 3 electrons.

Complete the diagram below to show the arrangement of the electrons in this **lithium** atom.

Use a cross to represent an electron.

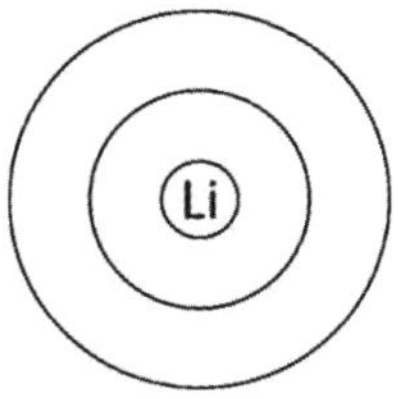

[1]

(d) Tick (✓) the name of the element that is found in the same **period** of the Periodic Table as the element lithium.

aluminium ☐

neon ☐

bromine ☐

sodium ☐

[1]

Self-assessment and reflective learning page
Periodic Table and salts

Tick (✓) the box which best shows how you feel about each statement.

Statement	I don't know	I need more practice	I understand
I understand that the structure of the Periodic Table is related to the atomic structure of the elements.			
I understand that the Periodic Table can be used to predict an element's structure and properties.			
I understand that the groups within the Periodic Table have trends in physical and chemical properties.			
I can use word equations to describe reactions.			
I can describe how to prepare some common salts by the reactions of metal carbonates with acids, and how to purify them using filtration and evaporation.			
Thinking and working scientifically			
I can use symbols and formulae to represent scientific ideas.			
I can make predictions of likely outcomes for a scientific inquiry based on scientific knowledge and understanding.			
I can decide what equipment is required to carry out an investigation or experiment and use it appropriately.			
I can describe trends and patterns in results, identifying any anomalous results and suggesting why results are anomalous.			
I can present and interpret results, and predict results between the data points collected.			

Mark for end of unit test _____ /20 marks

What went well in this topic?

What could you do better next time? What parts of the course could your teacher go through in a revision lesson which would support your improvement?

Teacher comment:

# End of Unit Test: Chemical bonding
Total = 20 marks

Name: .... Class: ....
Date: ....

1 (a) Identify the charge of an electron. Circle one answer.

    natural   positive   neutral   negative

[1]

(b) Draw a line to link each **type of bond** to its **description**.

Type of bond   **Description**

                                     attraction between oppositely charged icons

covalent   shared pair of atoms

ionic   shared pair of electrons

                                     attraction between similary charged electrons

[2]

(c) Explain how a zinc atom, Zn, becomes a zinc ion, Zn^{2+}.

....

....

[1]

2 The diagram below shows a molecule of ammonia.

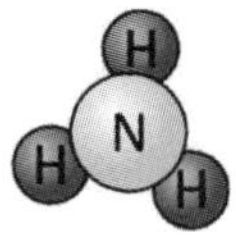

(a) (i) Write down the formula of a molecule of ammonia.

....

[1]

(ii) Describe how a molecule, such as ammonia, is formed.

....

....
[1]

(b) Explain why ammonia is a gas at room temperature.

....

....

....
[2]

3 George is investigating the density of some different objects.

The diagram below shows cube **A**.

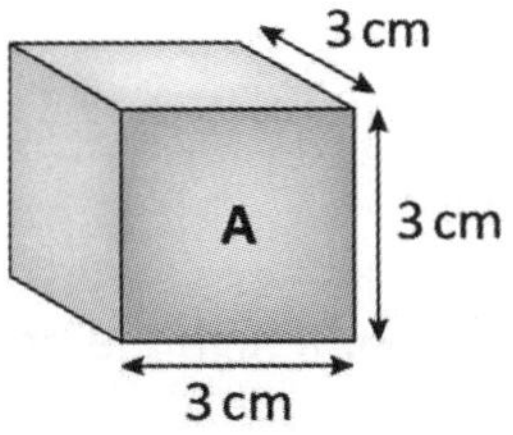

(a) Calculate the volume of cube **A**. Include the unit.

....
[2]

(b) Another cube labelled **B** has a volume a 4.0 cm³.

It has a mass of 3.6 g.

Calculate the density of cube **B**.

Use the equation:

$$\text{density} = \frac{\text{mass}}{\text{volume}}$$

.... [1]

26

(c) George wants to find the density of a type of rock. A pebble of the rock sinks in water.

(i) Describe how George could calculate the volume of the pebble.

Include the pieces of apparatus George should use.

....

....

....

....
[4]

(ii) Write down how George could adjust his method to find the volume of an irregularly shaped object that **floats** in water.

....

....
[1]

4 Diamond has some special properties. It has a high melting point and is a hard material that can be used to make the ends of drill bits.

drill bit

(a) State the type of bonding in diamond **and** the structure in diamond.

....

....
[2]

(b) Explain why diamond has these special properties.

....

....
[2]

Self-assessment and reflective learning page
Chemical bonding

Tick (✓) the box which best shows how you feel about each statement.

Statement	I don't know	I need more practice	I understand
I understand that a molecule is formed when two or more atoms join together chemically, through a covalent bond.			
I can describe a covalent bond as a bond made when a pair of electrons is shared by two atoms.			
I can describe an ion as an atom which has gained at least one electron to be negatively charged or lost at least one electron to be positively charged.			
I can describe an ionic bond as an attraction between a positively charged ion and a negatively charged ion.			
I can describe how the density of a substance relates to its mass in a defined volume.			
I can calculate and compare densities of solids, liquids and gases.			
I know that elements and compounds exist in structures (simple or giant), and this influences their physical properties.			
Thinking and working scientifically			
I can use symbols and formulae to represent scientific ideas.			
I can plan an investigation to obtain appropriate evidence.			
I can decide what equipment is required to carry out an investigation or experiment and describe how to use it appropriately.			
I can describe how to take appropriately accurate and precise measurements.			

Mark for end of unit test _____ /20 marks

What went well in this topic?

What could you do better next time? What parts of the course could your teacher go through in a revision lesson which would support your improvement?

Teacher comment:

End of Unit Test: Reactions and rates
Total = 20 marks

Name: .... Class: ....
Date: ....

1 Sadie investigates some reactions of magnesium and copper.

   When magnesium is heated in air using a Bunsen burner it burns with a brilliant white flame and forms magnesium oxide.

   (a) Name the piece of apparatus that Sadie should use to hold the magnesium in the Bunsen burner flame.

   ....
   [1]

   (b) (i) Write down one **hazard** of this investigation.

   ....
   [1]

   (ii) Write one way that Sadie could reduce the risk of an accident.

   ....
   ....
   [1]

   (c) Write a **word** equation for the reaction that happens when magnesium is burnt.

   ....
   [1]

   (d) (i) Calculate the mass of magnesium oxide formed when 2.4 g of magnesium reacts with 1.6 g of oxygen.

   ....
   [1]

   (ii) Predict the mass of oxygen that would react with 3.6 g of magnesium oxide.

   ....
   [1]

(e) The symbol equation for the reaction that happens when copper is heated is shown below.

2 Cu + O₂ → 2 CuO

Write a **word** equation for this reaction.

....

[1]

(f) Sadie wants to know the mass of oxygen and copper that react together. Describe how Sadie could carry out an experiment to find this out.

Include the equipment she should use, what she should do and the measurements she should take.

....

....

....

....

[4]

2 Lily has four metals: iron, magnesium, zinc and copper.

Lily adds each metal to different metal sulfate solutions and observes if a reaction takes place.

If a reaction occurs, she puts a tick in the table below. If there is no reaction, she puts a cross.

Metal sulfate solution / Metal	Iron	Magnesium	Zinc	Copper
iron sulfate		✓	✓	✗
magnesium sulfate	✗		✗	✗
zinc sulfate	✗	✓		✗
copper sulfate	✓	✓	✓	

(a) Name the type of reaction that takes place between magnesium and iron sulfate solution.

....

[1]

(b) Use the results in the table to identify the least reactive metal.

....

[1]

(c) Write a word equation for the reaction between **magnesium** and **iron sulfate**.

....

[1]

(d) Suggest why there is no reaction between **iron** and a solution of **magnesium sulfate**.

....

....

[1]

3 Scientists can change the rate of reaction by changing the conditions.

(a) Draw a line to link each change in conditions to how it affects the rate of a chemical reaction.

Change in conditions

- double the pressure of a gas
- halve the concentration of a solution
- increases the temperature from 20 °C to 30 °C

How it affects the rate of a chemical reaction

- increases the rate of reaction
- decrease the rate of reaction

[2]

Tim adds a piece of magnesium to a beaker of hydrochloric acid.

(b) (i) Name the products of this reaction.

.... and ....

[1]

(ii) Predict how using smaller pieces of magnesium would affect the rate of reaction. Explain your answer.

....

....

[2]

Self-assessment and reflective learning page
Reactions and rates

Tick (✓) the box which best shows how you feel about each statement.

Statement	I don't know	I need more practice	I understand
I can use word equations and symbol equations to describe reactions.			
I can identify examples of displacement reactions and predict products.			
I can describe the effects of concentration, surface area and temperature on the rate of reaction, and explain them using the particle model.			
I understand that in chemical reactions mass and energy are conserved.			
Thinking and working scientifically			
I can plan a range of investigations of different types to obtain appropriate evidence when testing hypotheses.			
I can decide what equipment is required to carry out an investigation or experiment and how to use it appropriately.			
I can describe how to carry out practical work safely.			
I can describe trends and patterns in results.			

Mark for end of unit test _____ /20 marks

What went well in this topic?

What could you do better next time? What parts of the course could your teacher go through in a revision lesson which would support your improvement?

Teacher comment:

# End of Unit Test: Energy and density
Total = 20 marks

Name: .... Class: ....
Date: ....

1   This question is about thermal energy and temperature.

   (a) Complete the following sentence using these terms.

   temperature   thermal energy

   When we measure ...., we are measuring the average

   .... of the particles in a substance.

   [1]

   (b) What quantity does a thermometer measure?

   ....

   [1]

2   This question is about conduction, convection and radiation.

   (a) Which type(s) of thermal energy transfer take place in a vacuum?

   ....

   [1]

   (b) Explain why the other type(s) of energy transfer **cannot** take place in a vacuum.

   ....

   ....

   [1]

3 Seren investigates floating and sinking.

The diagram shows where objects made from different materials end up after being placed in water.

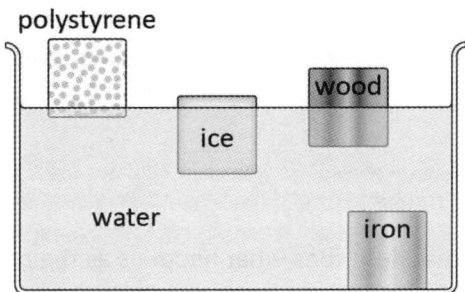

(a) Explain why ice floats on water.

....

[1]

(b) The table shows the densities of different materials.

Complete the table by identifying which material each object, **A**, **B** or **C**, is made from.

Write the letter of the object in the row of the material.

Material	Density (g/cm³)	Object
aluminium	2.7	
cork	0.2	
water	1.0	—
wood (cherry)	0.7	

[2]

4 Patrick investigates the transfer of thermal energy.

He uses the equipment in the diagram.

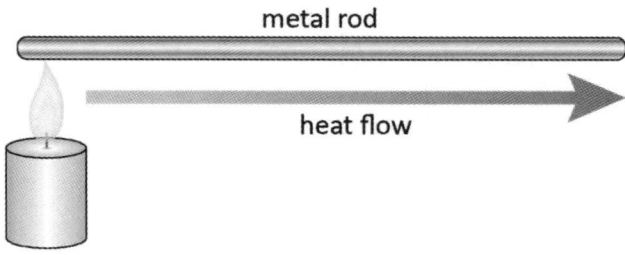

(a) Complete the sentence that describes what happens as the bar is heated. Circle the correct words.

Thermal energy is transferred from the **cooler / hotter** end of the bar to the **cooler / hotter** end of the bar.

[1]

(b) Patrick does not have a thermometer, but he does have some candle wax.

Describe how Patrick could show that thermal energy is transferred.

....

....

[2]

5 Ruby leaves a glass of water on a table indoors.

She forgets about it until a week later. Nobody else notices or touches the glass.

(a) Predict what Ruby will observe about the water when she collects the glass. Explain your answer.

....

....

[2]

(b) Ruby wants to investigate this scientifically.

She decides to leave another glass of water out.

Write down **one** thing that Ruby should do to make sure she gets a similar, but more scientific, result.

....

....

[1]

37

6 Joon investigates energy transfers.

He heats a pan of water, as shown in the diagram.

(a) Add **four** arrows to the diagram to show how the particles of water move within the pan.

[2]

(b) Write down **one** safety measure that Joon should take.

....

[1]

7 The diagram shows a rubber ball on a shelf.

The potential energy of the ball is 5 J.

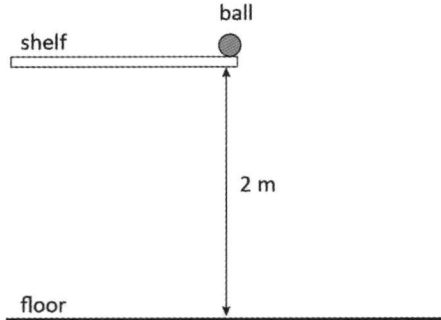

(a) The ball falls off the shelf, towards the ground.

Just before it reaches the ground, the ball has 5 J of kinetic energy.

How much potential energy does the ball have now? Explain your answer.

....J

because: ....

....

[2]

(b) The ball then hits the ground and bounces.

It rises upwards but only reaches a height of 1.4 m (which is lower than the shelf).

The total energy stored by the ball is now 3.5 J.

(i) Calculate the amount of energy wasted by the ball.

.... J
[1]

(ii) What has happened to this wasted energy?

....

....

....
[1]

Self-assessment and reflective learning page
Energy and density

Tick (✓) the box which best shows how you feel about each statement.

Statement	I don't know	I need more practice	I understand
I can use density to explain why objects float or sink in water.			
I understand the difference between heat and temperature.			
I know that energy is conserved, so that it cannot be created or destroyed.			
I know that thermal energy always transfers from hotter things to colder things (heat dissipation).			
I understand thermal transfer by conduction, convection and radiation.			
I can explain cooling by evaporation.			
Thinking and working scientifically			
I can sort materials through testing and observation.			
I can evaluate investigations and suggest improvements.			
I can make predictions based on scientific knowledge and understanding.			
I can plan an investigation.			
I can identify and control risks.			
I can make conclusions by interpreting results.			

Mark for end of unit test _____ /20 marks

What went well in this topic?

What could you do better next time? What parts of the course could your teacher go through in a revision lesson which would support your improvement?

Teacher comment:

# End of Unit Test: Electricity and sound
Total = 20 marks

Name: .... Class: ....

Date: ....

1 Mia records the oscilloscope traces for two sound waves.

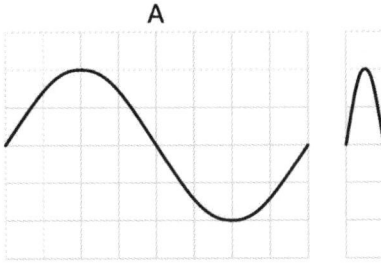
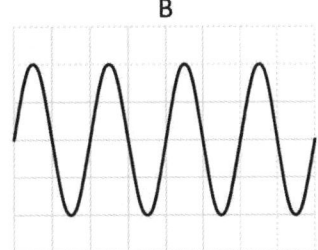

(a) Which sound will have the higher pitch?

....

[1]

(b) Which equation shows how wave B is related to wave A? Tick (✓) **one** answer.

frequency of B = frequency of A ☐

frequency of B = 2 × frequency of A ☐

frequency of B = 3 × frequency of A ☐

frequency of B = 4 × frequency of A ☐

[1]

2 Chen investigates a parallel circuit.

This is the circuit diagram.

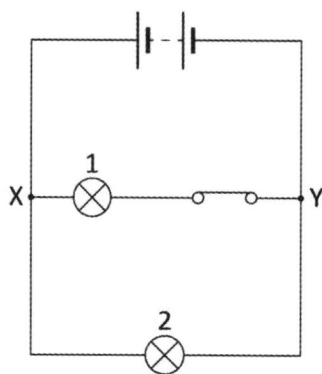

(a) Describe what happens to the electric current at the junction marked **X**.

....

[1]

(b) Chen wants to turn each lamp on or off separately.

He needs to add one more component to do this.

Name the component and describe where it should be placed.

....

....

[1]

3 Timi uses an ammeter to measure the current in a lamp.

After leaving the lamp on for a few minutes, she notices two things:

(i) the lamp gives off more heat

(ii) the ammeter reading decreases.

(a) What happens to current in a circuit if resistance increases? Circle **one** answer.

        decreases   **stays the same**   **increases**

[1]

(b) Write down what happens to the resistance of the lamp when it is left on.

....

[1]

4 The diagram shows oscilloscope traces for two interfering sound waves.

The crest of one wave meets the crest of the other wave.

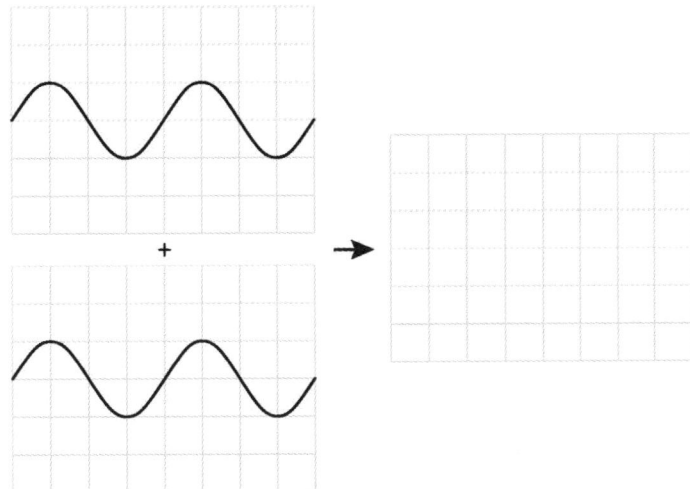

(a) Sketch the wave that is produced on the right-hand side of the diagram.

[2]

(b) How loud will the resulting wave be compared to each interfering wave? Write a short description.

....

[1]

5 Hamid sets up the following circuit.

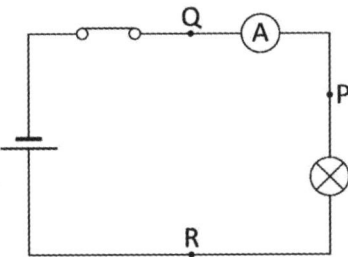

The reading on the ammeter is 0.5 A.

(a) Hamid adds a lamp at the position P.

Describe what will happen to the ammeter reading.

....

[1]

(b) Now Hamid connects another lamp between points Q and R.

(i) Is this new lamp in series or in parallel with the first two lamps?

....

[1]

(ii) Predict how bright the new lamp will shine compared to the first two lamps.

....
[1]

(c) What can Hamid do to make **all three** lamps shine more brightly?

....
[1]

6 Dilpa investigates the resistance of a variable resistor.

Look at the diagram of the circuit that Dilpa uses.

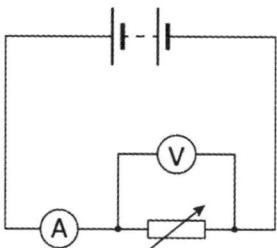

After changing the variable resistor, Dilpa measures the voltage and current.

She calculates the resistance each time using the formula:

resistance = $\dfrac{\text{voltage}}{\text{current}}$

The table shows Dilpa's results.

Voltage (V)	Current (A)	Resistance (Ω)
6.0	1.5	4.0
6.0	1.2	5.0
6.0	0.75	
6.0	0.50	.

(a) Calculate the missing values of resistance and write them in the table.

[2]

(b) Describe how Dilpa could improve the reliability of the results.

....
[1]

7 Omar is testing cells using a circuit containing a voltmeter and a lamp.

All the cells are meant to have a voltage of 2.0 V.

The voltmeter uses a pointer on a scale.

(a) Look at the voltmeter in the diagram.

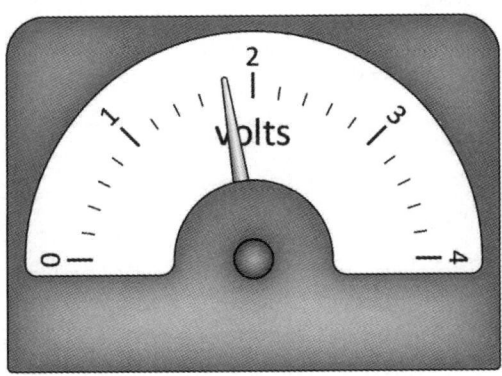

Write down the reading it shows.

....
[1]

(b) Complete the diagram of the circuit shown below.

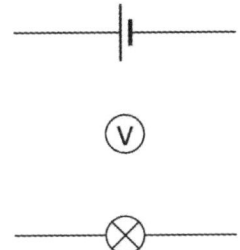

[1]

(c) The table shows Omar's results for 5 different cells.

Cell	1	2	3	4	5
Voltage (V)	1.9	2.0	2.1	1.4	1.9

The result for cell 4 is anomalous.

Suggest **two** possible reasons for this reading.

1 ....

2 ....

[2]

Self-assessment and reflective learning page
Electricity and sound

Tick (✓) the box which best shows how you feel about each statement.

Statement	I don't know	I need more practice	I understand
I can draw and interpret waveforms.			
I understand the link between loudness and amplitude of sound waves.			
I understand the link between pitch and frequency of sound waves.			
I can use waveforms to show how sound waves interact with each other.			
I can describe how current divides in parallel circuits.			
I know how to measure current and voltage in series and parallel circuits.			
I can describe the effects on current and voltage of adding cells and lamps.			
I can calculate resistance.			
I can describe how resistance affects current.			
I can use symbols to represent circuits.			
I can make and compare different circuits.			
Thinking and working scientifically			
I can interpret the results of an investigation.			
I can decide what equipment is needed for an investigation and how to use it.			
I can predict the outcome of an investigation based on scientific knowledge.			
I can evaluate investigations and suggest improvements.			
I can use symbols and formulae to represent scientific ideas.			
I can record measurements in a suitable form.			
I can describe how to improve the reliability of results.			
I can make accurate measurements.			
I can identify anomalous results and suggest explanations for them.			

Mark for end of unit test _____ /20 marks

What went well in this topic?

What could you do better next time? What parts of the course could your teacher go through in a revision lesson which would support your improvement?

Teacher comment:

# End of Unit Test: Earth and space
Total = 20 marks

Name: ………………………………………. Class: …………………………………………..
Date: ………………………………………..

1   Scientists think that the Moon formed around 100 million years after the planets formed in the Solar System.

   (a)   Which hypothesis seems most likely to explain how the Moon formed? Circle **one** answer.

   capture hypothesis   **co-formation hypothesis**   **collision hypothesis**

   [1]

   (b)   Describe **one** piece of evidence that supports this hypothesis.

   ……
   [1]

2   This question is about the carbon cycle.

   (a)   The flow chart shows how carbon from the air is incorporated by decomposers.

   Complete the flow chart by naming the unlabelled processes.

   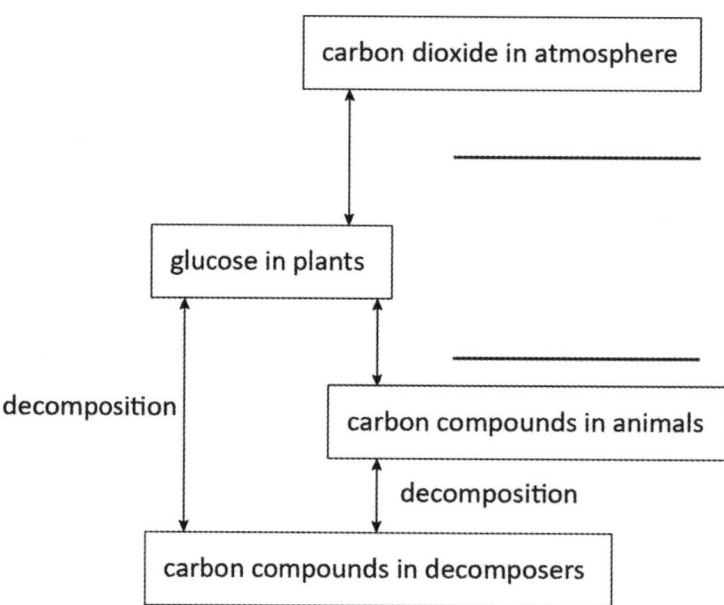

   [1]

The processes in the flow chart **reduce** the level of carbon dioxide in the atmosphere.

The table shows the mean amount of carbon dioxide in the atmosphere over some of the last 65 years.

Year	1960	1970	1980	1990	2000	2010	2020	2023 (estimate)
mean level of carbon dioxide in atmosphere (parts per million) for that year	315	325	338	353	369	388	412	420

(b) Describe the trend shown by the data.

....
[1]

(c) Describe **two** processes that need to be added to the flow chart in part (a) to explain the trend in the data.

....

....

....
[2]

3 Teams of astronomers across the world identify and track the movements of asteroids.

They use large, expensive telescopes on the ground and in space.

(b) Describe the main reason why the movements of asteroids must be tracked.

....
[1]

(b) Suni is planning a science project about asteroids.

Decide whether Suni should make her own observations or use results from the astronomers' investigations. Explain your answer.

....

....
[2]

(c) Asteroids have collided with Earth in the past.

Describe **one** effect on Earth of a collision.

....

....

[1]

4 This question is about the changing climate.

(a) Explain why climate change is leading to rising sea levels.

....

[2]

(b) Average sea level is increasing by 3.4 mm each year.

Calculate the change in sea level over 50 years, in centimetres.

....cm

[1]

5 This question is about nebulae.

(a) What causes a nebula to form? Tick (✓) **two** statements.

An asteroid collides with a planet.	
A black hole emits gases.	
A very large star explodes.	
Interstellar gases are pulled together by gravity.	

[1]

(b) Which component of a nebula is most important for the formation of new stars? Explain your answer.

....

....

[1]

6 Read the following short article about Iceland.

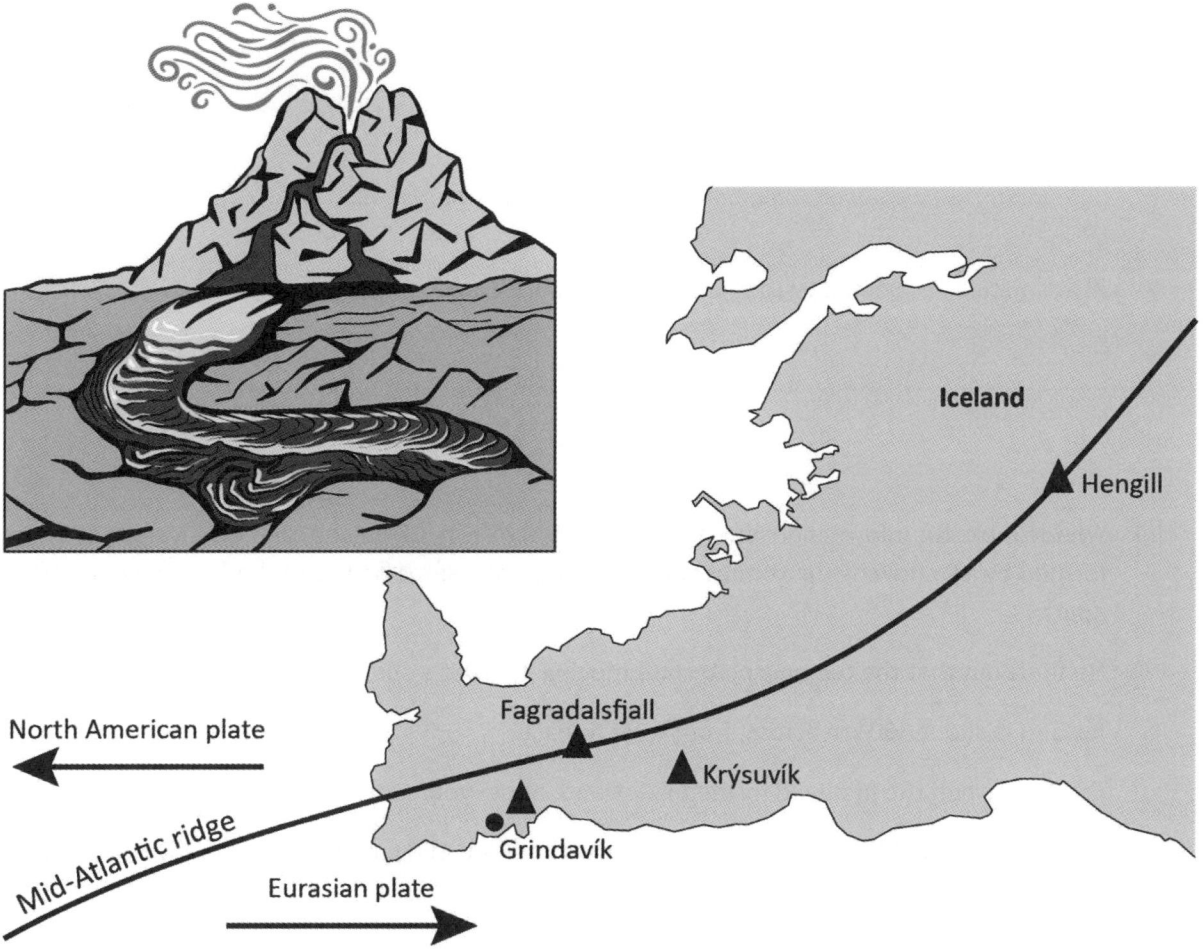

Date: 18 December 2023

Location: 4 km north of Grindavik, a town in Iceland

At 10.17 pm, a fissure (crack) in the Earth's crust opened and hot, liquid rock was thrown up to 100 m into the air. Lava flowed outwards from the fissure.

This followed 6 weeks of earthquakes in the area, which led to the evacuation of Grindavik.

The map shows the locations of recent eruptions and the Mid-Atlantic ridge, which follows the line where the North American and Eurasian tectonic plates meet.

(a) The map shows that in Iceland, two tectonic plates are moving apart.

Explain what causes tectonic plates to move.

....

....

[1]

(b) The article includes **two** pieces of evidence for the movement of tectonic plates.

Write a short explanation of each piece of evidence.

1 ....

....

2 ....

....

[2]

(c) Ayinde looks for information about the geological history of Iceland, and discovers that it was formed by lava flowing up from between the tectonic plates on the seabed, as they moved apart.

He finds out that the tectonic plates are moving apart at a speed of about 2 cm per year.

Iceland is about 500 km across, from west to east.

Estimate when the first land formed in Iceland. Show your working.

(First, convert 500 km to cm. Remember that there are 100 cm in 1 m, and 1000 m in 1 km.)

Age ....years

[2]

Self-assessment and reflective learning page
Earth and space

Tick (✓) the box which best shows how you feel about each statement.

Statement	I don't know	I need more practice	I understand
I can describe the evidence for the collision theory of how the Moon formed.			
I can describe the carbon cycle.			
I can describe how an asteroid collision affected the Earth.			
I can describe the impacts of climate change.			
I know that nebulae are clouds of dust and gas, where new stars can form.			
I can explain how convection currents cause the movement of tectonic plates.			
I can explain the evidence for the movement of tectonic plates.			
Thinking and working scientifically			
I can describe models and discuss their strengths and limitations.			
I understand that models can change.			
I can describe trends in results.			
I can decide whether to use evidence from first-hand investigations or from secondary sources.			
I can show how evidence supports or refutes a prediction or idea.			

Mark for end of unit test _____ /20 marks

What went well in this topic?

What could you do better next time? What parts of the course could your teacher go through in a revision lesson which would support your improvement?

Teacher comment:

End of Year Test 1
Total = 50 marks

Name: .... Class: ....
Date: ....

1 In a chemical reaction, reactants mix and react to form new products.

(a) The formulae of the substances change during the reaction. However, two other quantities stay the same. Which quantities? Tick (✓) **two** answers.

  A temperature of substances ☐

  B temperature of surroundings ☐

  C total energy of substances ☐

  D total energy of substances + surroundings ☐

  E total mass of substances ☐

[1]

(b) Complete the word equation for the following chemical reaction.

  calcium + ... → calcium hydroxide + hydrogen

[1]

2 The diagram shows an electric circuit and a hot water heating circuit for a house.

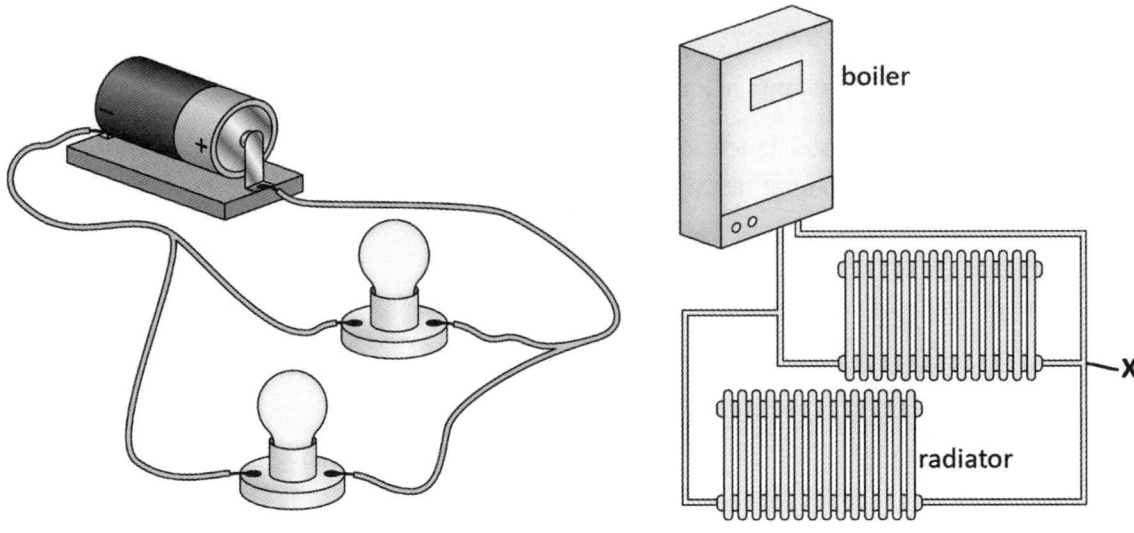

The hot water circuit is an **analogy** for an electric circuit. This means that the flow of water around the hot water circuit can be used as a model for the current in an electric circuit.

(a) Which component in the electric circuit is represented by the radiator?

....

[1]

(b) What happens to the water current that flows into the junction marked **X**? Use this to explain what happens to the current at a junction in a parallel electric circuit.

....

....

[2]

3 The diagram shows how a particular substance moves through a plant.

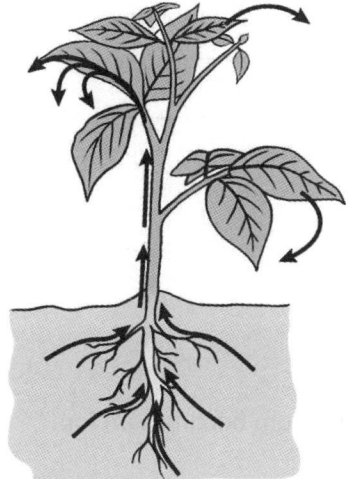

(a) Write down the substance that moves through a plant in this way.

....
[1]

(b) This substance is lost from the plant through stomata on the surface of leaves.

State **two** types of weather conditions that speed up this loss.

1 ....

2 ....

[2]

4 Ashwin observes the night sky using a telescope.

He sees many bright points of light (stars) as well as some areas between stars that look like coloured mist. The star map he uses calls these areas nebulae.

(a) Describe the contents of a nebula.

....

....

....

[1]

(b) Over millions of years, some nebulae will increase in size while others will reduce in size.

Explain the events taking place in these two types of nebulae.

1 increase: ....

2 decrease: ....

[2]

5 Potassium is a metallic element.

(a) Describe how an atom of potassium becomes an ion.

....

....

[1]

(b) Akuba has the following list of particles:

$$Na^+ \quad Mg^{2+} \quad Cl^- \quad Ne$$

Which of these particles will form an ionic bond with potassium? Explain your answer.

Particle: ....

because ....

....

[2]

6 Bonnie and Craig are expecting their first baby.

As soon as she knows she is pregnant, Bonnie stops smoking.

(a) How can harmful substances pass from a mother to a developing foetus?

....

[1]

Bonnie asks Craig to stop smoking as well, but he says that him stopping smoking will not affect the baby's health.

(b) What should Craig do? Write a short scientific explanation for Bonnie and Craig.

....

....

[2]

7 Iwan investigates sound waves produced by two loudspeakers.

He uses the equipment shown in the diagram to produce continuous sounds of the same amplitude and frequency.

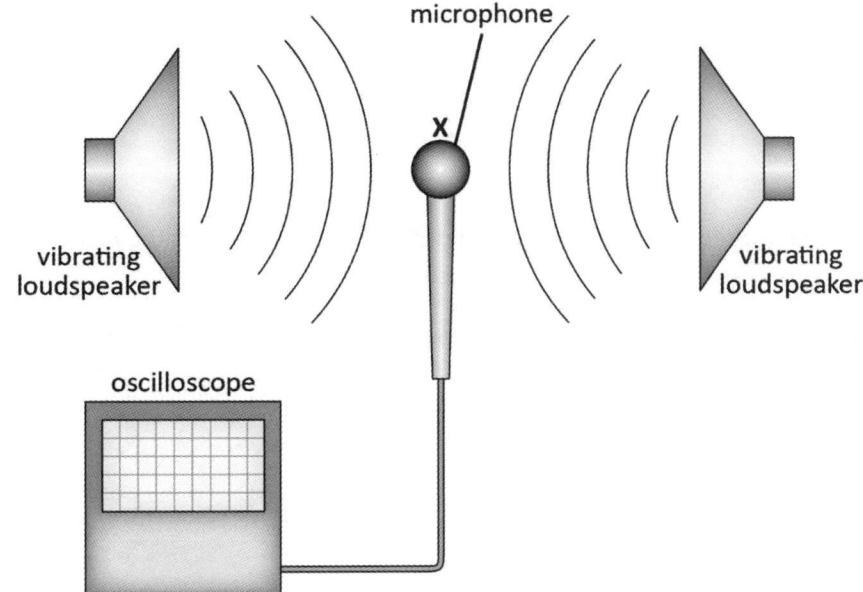

Iwan records the trace from the oscilloscope when the microphone is placed at point **X**.

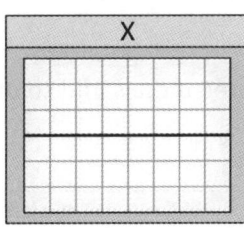

(a) Explain what is happening to the sound waves at point **X**.

....

....
[1]

(b) Iwan thinks the result at **X** might be an anomaly.

Suggest how Iwan could adjust his investigation to show that **X** is an accurate result, not an anomaly. Include a prediction of the outcome of the change.

....

....

....
[2]

8 This question is about inheritance.

   (a) Complete the following sentence to describe a gene.

   A gene is a section of .... that controls the development of a

   specific ....

   [1]

   (b) Some people are red–green colour vision deficient, meaning that they cannot distinguish between certain shades of red and green.

   Scientists think that this condition is caused by differences in a single gene, which they have identified.

   Explain how this idea could be investigated.

   ....

   ....

   ....

   [2]

9 Amara investigates the reactions of three metals and solutions of their salts.

   The table shows the results.

Metal added to solution	Solution		
	Aluminium sulfate	Calcium sulfate	Copper sulfate
aluminium	✗	✗	✓
calcium	✓	✗	✓
copper	✗	✗	✗

   (a) Name the type of reaction that Amara used.

   ....

   [1]

   (b) Use the results to place the three metals in a reactivity series with the most reactive metal first.

   most reactive ....

   ....

   least reactive ....

   [1]

(c) Write down **one** safety precaution that Amara should use.

....
[1]

10 Harjeet investigates the effect of heating a metal bar.

He uses the equipment shown in the diagram. The bar is held in position by clamps (not shown).

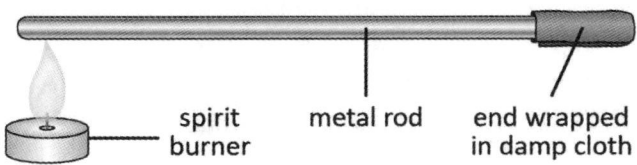

After a few minutes, Harjeet uses a probe to measure the temperature of the rod at different distances from the burner.

The table shows his results.

	Distance along rod from burner (cm)						
	0	5	10	15	20	25	30
Temperature (°C)	70	60	53	47	42	38	35

(a) Choose the correct definition for temperature. Tick (✓) **one** statement.

A average amount of thermal energy stored ☐

B total amount of thermal energy stored ☐

C average amount of thermal energy transferred ☐

D total amount of thermal energy transferred ☐

[1]

(b) Describe and explain the results.

....

....
[2]

(c) Harjeet puts out the burner flame. He observes that the cloth has dried out.

Suggest what has happened to the water in the cloth.

....
[1]

11 The diagram shows the carbon cycle. Some arrows and labels are missing.

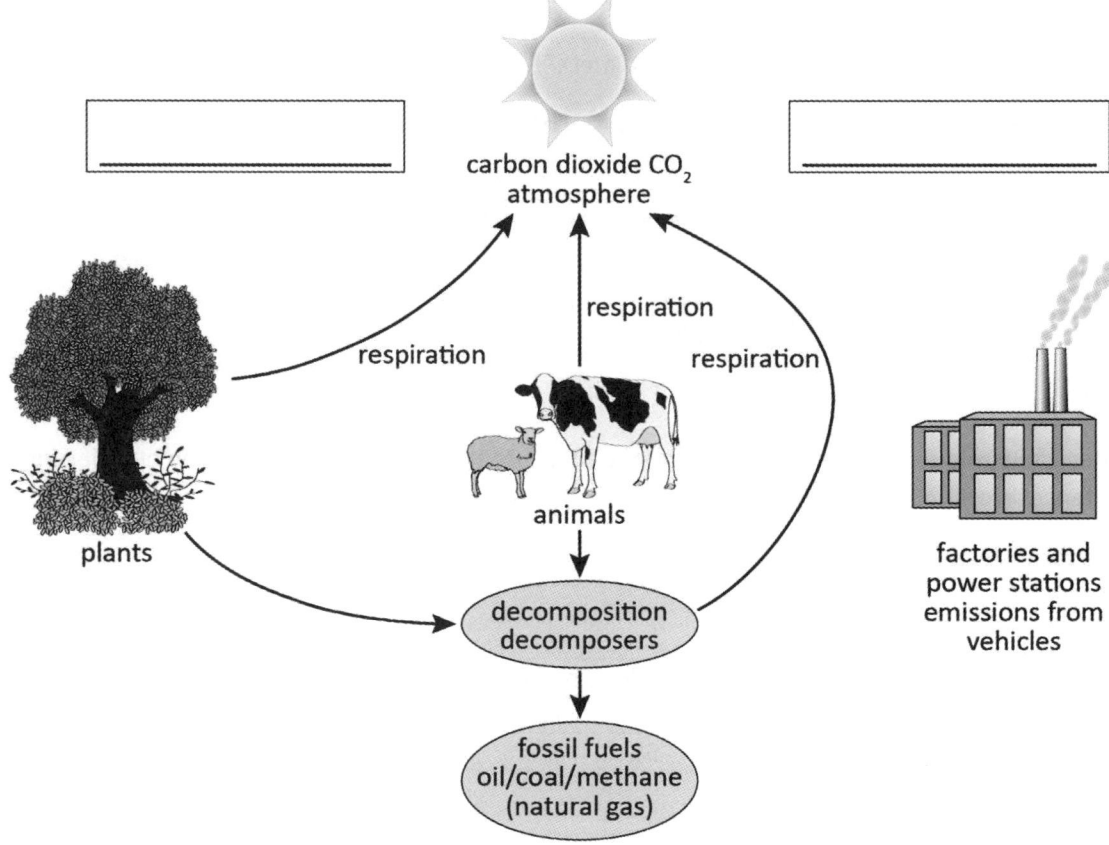

(a) Complete the diagram by adding **three** arrows and **two** labels.

[2]

The table lists some actions that some countries took during 2023 that could have an impact on climate change.

	Reduces carbon dioxide emissions over time?
Build coastal defences to reduce flooding from the sea.	☐
Build more wind farms to produce electricity.	☐
Issue licences to extract natural gas from new areas for fuel.	☐
Increase amount of insulation required in new houses.	☐
Open new coal mines as a source of fuel.	☐

(b) In the table, tick (✓) the actions that will help to **reduce** the amount of carbon dioxide produced by human activities, over time.

[2]

The graph shows the measured and predicted change in average sea level from the year 2000 until 2100. The prediction is based on carbon dioxide levels in the atmosphere continuing to increase.

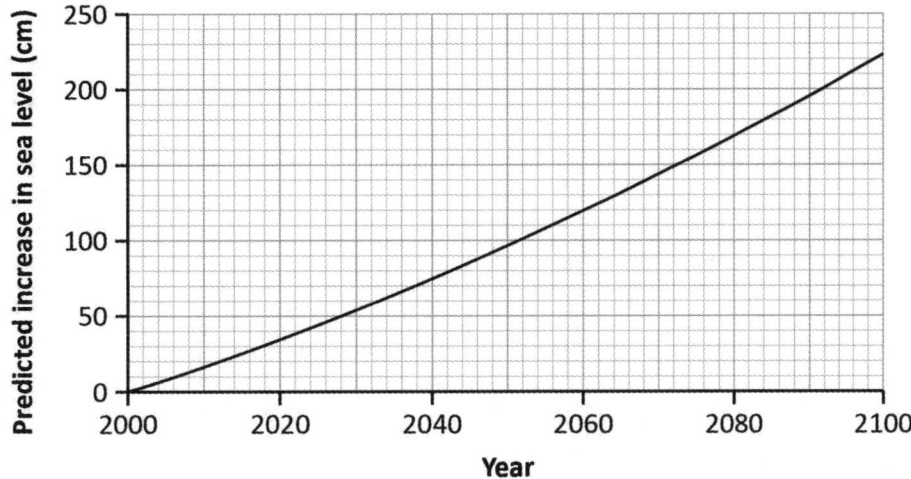

In the year 2000, a coastal city already had sea walls that could cope with a 40 cm rise in average sea level. Planners intend to increase the height of sea walls by another 60 cm by the year 2060.

(c) State the sea wall height change you would advise. Give **two** reasons for your answer.

....cm

....

....

....

....

[2]

12 The diagram shows the Periodic Table. The elements in Group 2 are shaded.

1	2											3	4	5	6	7	0
						H hydrogen 1	Hydrogen has the atomic number 1										He helium 2
Li lithium 3	Be beryllium 4											B boron 5	C carbon 6	N nitrogen 7	O oxygen 8	F flourine 9	Ne neon 10
Na sodium 11	Mg magnesium 12											Al aluminium 13	Si silicon 14	P phosphorus 15	S sulfur 16	Cl chlorine 17	Ar argon 18
K potassium 19	Ca calcium 20	Sc scandium 21	Ti titanium 22	V vanadium 23	Cr chromium 24	Mn magnesium 25	Fe iron 26	Co cobalt 27	Ni nickel 28	Cu copper 29	Zn zinc 30	Ga gallium 31	Ge germanium 32	As arsenic 33	Se selenium 34	Br bromine 25	Kr krypton 36
Rb rubidium 37	Sr strontium 38	Y yttrium 39	Zr zirconium 40	Nb niobium 41	Mo molybdenum 42	Tc technetium 43	Ru ruthenium 44	Rh rhodium 45	Pd palladium 46	Ag silver 47	Cd cadmium 48	In indium 49	Sn tin 50	Sb antimony 51	Te tellurium 52	I iodine 53	Xe xenon 54
Cs caesium 55	Ba barium 56	La* lanthanum 57	Hf hafnium 72	Ta tantalum 73	W tungsten 74	Re rhenium 75	Os osmium 76	Ir iridium 77	Pt platinum 78	Au gold 79	Hg mercury 80	Tl thallium 81	Pb lead 82	Bi bismuth 83	Po polonium 84	At astatine 85	Rn radon 86
Fr francium 87	Ra radium 88	Ac** actinium 89	Rf rutherfordium 104	Db dubnium 105	Sg seaborgium 106	Bh bohrium 107	Hs hassium 108	Mt meitnerium 109	Ds darmstadium 110	Rg roentgenium 111							

(a) How many electrons are found in the outer shell of atoms of the elements in Group 2?

....

[1]

The elements in Group 2 show similar patterns of properties to the elements in Group 1.

(b) Lula wants to investigate the melting points of the elements in Group 2.

Write a hypothesis for Lula to test. Circle your choice.

Going down Group 2, the melting point of the elements **decreases / stays the same / increases**.

[1]

The table shows Lula's results.

Element	Melting point (°C)			
	Test 1	Test 2	Test 3	Mean
Be, beryllium	1290	1280	1270	1280
Mg, magnesium	640	660	650	650
Ca, calcium	835	830	855	....
Sr, strontium	770	755	770	....
Ba, barium	715	730	730	725

(c) Calculate the mean melting points for calcium and strontium. Write them in the table.

[2]

(d) Determine whether these results do or do not support the hypothesis from part (b).

....

....
[1]

(e) All the melting points are high.

Suggest a reason for this.

....

....
[1]

13 Emma investigates the growth of tomato plants over several weeks.

She puts four groups of seeds in soil in separate pots, **A–D**.

The seeds in pot **A** are given water only.

The seeds in pots **B–D** are given water and different mixtures of minerals.

The table shows Emma's results after 8 weeks.

Pot	Amount of water given to plants daily (ml)	Mineral mixture given to plants	Observations of plants	
			Mean height (cm)	Leaf colour
A	400	none	41	yellow
B	400	mixture 1	124	green
C	400	mixture 2	83	green
D	400	mixture 3	118	yellow

(a) (i) Which mineral do plants in pots **A** and **D** lack? Circle **one**.

magnesium   nitrogen   phosphorus   potassium

[1]

(ii) The word equation for photosynthesis is:

carbon dioxide + water → glucose + oxygen

Explain how a lack of the mineral named in part (i) affects this reaction.

....

....
[1]

(b) Calculate how many times higher the plants in pot **B** grew compared to the pot that had no added minerals.

....
[1]

(c) Describe **two** ways that Emma could obtain more reliable results.

1 ....

....

2 ....

....
[2]

14 Mika investigates the resistance of an electrical component.

He draws the following circuit diagram.

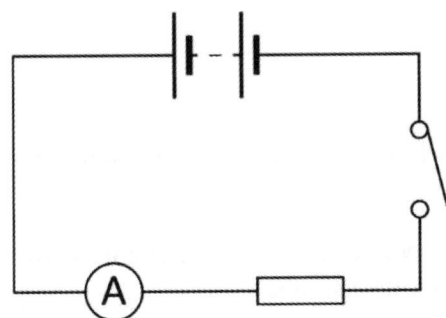

(a) One component is missing from Mika's diagram.

Add this component to the correct place in the diagram.

[1]

The table shows Mika's results.

Test	Temperature (°C)		
	Current (A)	Voltage (V)	Resistance (Ω)
1	0.42	4.6	11.0
2	0.38	4.3	11.3
3	0.42	4.5	

(b) Calculate the missing value for resistance. Add it to the table.

Resistance = ... Ω

[1]

(c) Calculate the mean resistance.

Mean resistance = ... Ω

[1]

End of Year Test 2
Total = 50 marks

Name: .... Class: ....

Date: ....

1 Look at the table describing cells involved in human reproduction.

Cell	Number of chromosomes	Sex chromosome(s)
egg cell	23	X
sperm cell	23	X or Y
fertilised egg cell	46	XX or XY

(a) Describe how the egg and sperm cells are involved in fertilisation.

....

[1]

(b) Why is it important that the sperm cell can have either an X or Y sex chromosome?

....

[1]

2 The Alvarez hypothesis suggests that an asteroid collided with Earth around 66 million years ago.

Scientists think that this event caused a mass extinction of plants and animals.

(a) What causes an asteroid to be pulled towards Earth?

....

[1]

(b) Which of the following effects of an asteroid collision may cause climate change?

Tick (✓) **one** box.

Blasts of heat directly from the impact. ☐

Clouds of dust and gas thrown high into Earth's atmosphere. ☐

Sudden gusts of wind at extreme speeds. ☐

Tsunamis if the impact is in or near an ocean. ☐

[1]

(c) Explain what occurs in a mass extinction event.

....
[1]

3 Kato is going for a ride in a hot air balloon, like the one shown in the diagram.

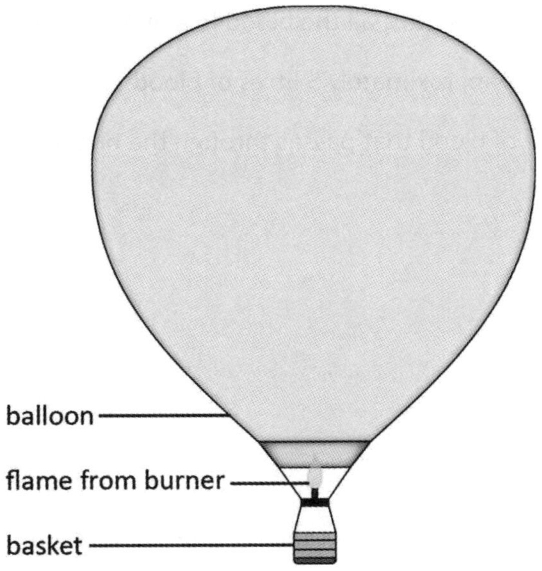

balloon
flame from burner
basket

Air inside the balloon is heated by the burner until the balloon is full of warm air.

(a) (i) Name the process in which thermal energy transfers from the heat of the burner to the air in the balloon. Circle **one** answer.

      conduction   **convection**   **radiation**

[1]

(ii) Add arrows on the diagram to show how warmer and colder air move inside the balloon.

[1]

(b) Kato notes that:

energy transferred from burner = energy transferred to air inside balloon + energy transferred to surroundings

What important principle does this equation describe?

....
[1]

4   This question is about the human excretory system.

    (a)   Name the main waste substance removed from the blood by the kidneys.

    ....
    [1]

    (b)   Approximately every 30 minutes, all the blood in a human body passes through the kidneys.

    An average human has approximately 5 litres of blood.

    Calculate the volume of blood that passes through the human body in 1 day. Show your working.

    ....litres
    [2]

5   The diagram shows the outermost electron shell of a bromine atom.

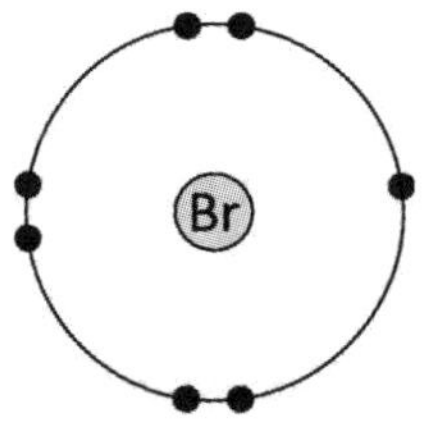

    Two atoms of bromine will join to form a molecule, Br_2.

    (a)   Identify the type of bond in a molecule of bromine. Circle **one** answer.

    covalent   **ionic**   **metallic**

    [1]

    (b)   Draw a dot-and-cross diagram to show the bond in a molecule of bromine.

    [2]

6 The dodo was a bird that lived on the island of Mauritius.

Dodos had plenty of food that was easy to find, and they had few predators.

Over millions of years, they adapted to the environment by:

- increasing in size
- losing the ability to fly.

(a) Name this process of adaptation.

....
[1]

(b) In 1507, humans arrived on Mauritius with farm animals.

By 1700, the dodo was extinct.

Suggest **one** reason for this extinction.

....

....
[1]

7 Anna investigates whether objects made of different materials will float or sink.

She uses the equipment shown in the diagram.

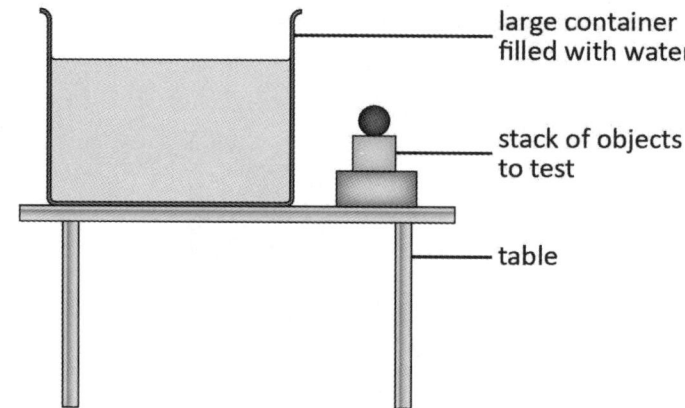

(a) Suggest **one** change Anna should make to reduce a risk or hazard. Explain your answer.

Change: ....

Explanation:..

[2]

The table shows Anna's results.

Object	Material	Sink or float?
cube	polystyrene	float
rectangular block	balsa wood	float
ball bearing	steel	sink

(b) Explain why the ball bearing sinks.

....

[1]

8 Kwame investigates the density of olive oil.

(a) Describe the equipment Kwame can use to determine the density of the oil.

....

[1]

(b) Kwame finds that 50.0 cm³ of olive oil has a mass of 46.0 g.

Calculate the density of olive oil. Show your working and include the units in your answer.

....
[2]

9 Susana investigates the sound waves produced by two different tuning forks, **P** and **Q**.

She uses an oscilloscope to show the trace of each wave.

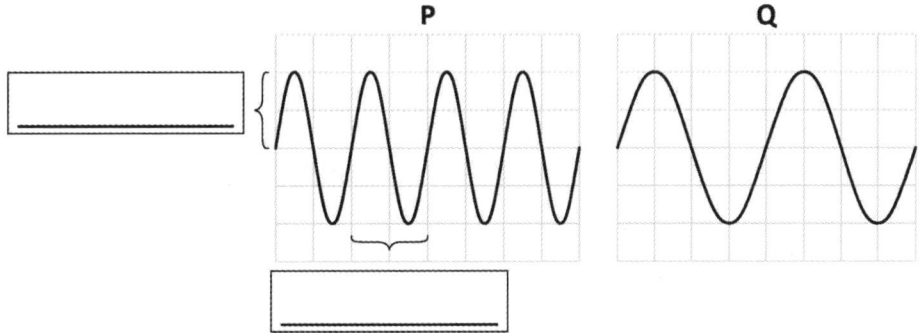

(a) Complete the labels for wave trace **P**.

[1]

(b) Compare the pitches of the tuning forks. Circle the correct term.

The pitch of **Q** is **the same as / lower than / higher** the pitch of **P**.

[1]

10 Calcium carbonate reacts with dilute hydrochloric acid.

Peter investigates how the rate of this reaction is affected by the size of pieces of calcium carbonate used. He uses the same mass of reactants in each test.

(a) Complete the word equation for this reaction.

calcium carbonate + hydrochloric acid →

.... + ... + water

[1]

Peter collects the gas produced using a gas syringe.

The diagram shows the gas syringe containing some gas.

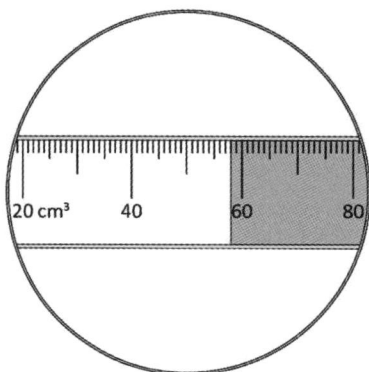

(b) Write down the volume of gas shown. Include the unit.

volume ....

[1]

Peter's results are shown in the table.

Time (s)	Volume of gas produced (cm³)	
	For large pieces of calcium carbonate	For small pieces of calcium carbonate
0	0	0
10	8	15
20	15	29
30	22	43
40	28	58
50	35	68
60	42	74
70	48	76
80	54	76

(c) Explain Peter's results using the particle model.

....

....

[2]

11 Xin Yi investigated the eye colours of students in her class.

The graph shows her results.

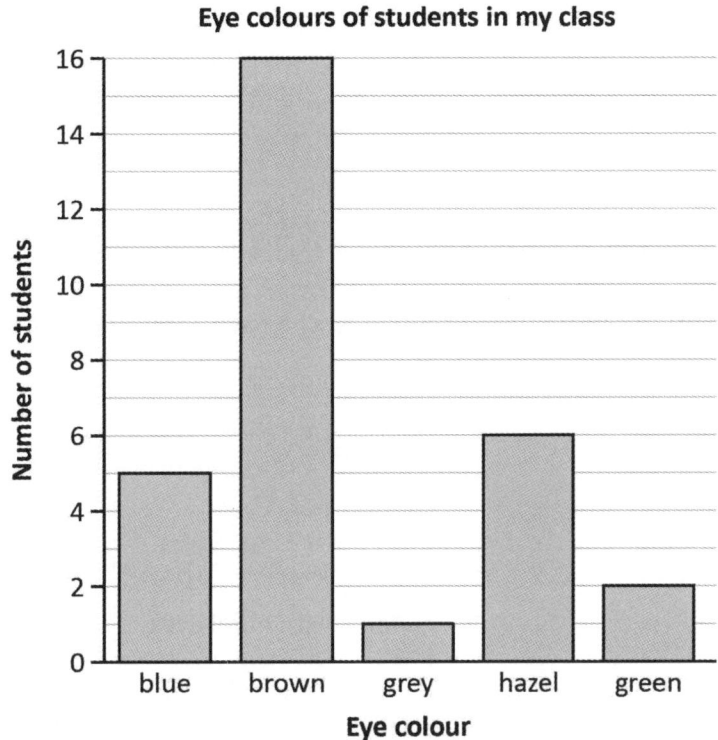

(a) What type of variation do these results show? Circle **one** answer.

        anomalous   **continuous**   **discontinuous**

[1]

(b) Deduce whether this variation is caused by environment, inheritance or a mixture of both.

....

[1]

(c) The 'sample size' is the total number of students in Xin Yi's class.

(i) Calculate the sample size.

sample size ....

[1]

(ii) Calculate the percentage of students who have hazel eyes. Show your working.

students with hazel eyes ....

[2]

(d) Xin Yi's results may not be reliable as a guide to eye colour for the general population.

Suggest how Xin Yi could adapt her investigation to produce more reliable estimates.

....

[2]

12 This question is about the effects of adding lamps or cells to electric circuits.

(a) Describe what happens to the brightness of a lamp in a series circuit when:

(i) the current through the lamp decreases

....

[1]

(ii) a cell is added to the circuit in series.

....

[1]

(b) Skye investigates the effects of adding components on electric circuits.

Skye has:

- connecting wires
- switch
- ammeter
- voltmeter
- 3 cells
- 3 lamps

The first circuit Skye makes contains a switch, 1 cell and 1 lamp.

He is not sure where to connect the ammeter or voltmeter.

(i) Complete the circuit diagram below.

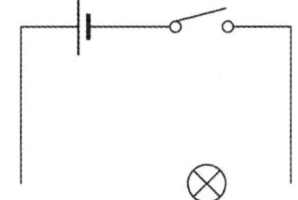

[2]

Skye measures the current in the circuit and the voltage across the lamp.

He then adds lamps in series one at a time and repeats his measurements.

The table shows his results.

Number of lamps in series	Current (A)	Voltage across lamp 1 (V)
1	0.30	1.5
2	0.15	0.75

(ii) Describe **two** ways in which Skye could change the circuit to make 2 lamps shine as brightly as 1 lamp on its own.

....

....

....

[2]

13 Rhian wants to make a salt by reacting magnesium carbonate with hydrochloric acid.

(a) Name the salt formed when magnesium carbonate reacts with hydrochloric acid.

....

[1]

(b) Describe how Rhian could test the gas produced in the reaction. Include the expected result in your description.

....

....

....

[2]

(c) Rhian purifies the salt solution using the following steps:

- separate the products from unreacted magnesium carbonate by filtration
- place the salt solution in an evaporating basin and warm it gently until all the water has evaporated.

The result is a small amount of powdery salt.

Explain how Rhian could improve this process to get larger crystals of the salt.

....

....

[1]

14 This question is about tectonic plates.

(a) Which of the following causes tectonic plates to move? Tick (✓) **one** box.

changes in the Earth's magnetic field ☐

conduction of thermal energy through the Earth's crust ☐

convection currents in the mantle ☐

radiation from the Sun ☐

[1]

The table lists fossilised organisms that have been found in different places on Earth.

Organism	Where found	Range of time when alive on Earth
Cygnonathus (land reptile)	South America, Africa	247 to 228 million years ago
Glossopteris (land plant)	South America, Africa, Southern Asia, Antarctica, Australia	300 to 200 million years ago
Minmi (land dinosaur)	Australia	121 to 112 million years ago
Trilobite (sea animal)	South America, Africa, Southern Asia, Australia	520 to 250 million years ago

Scientists think that several continents were joined together until 180 million years ago (see diagram below). After this time, the movement of tectonic plates caused continents to move apart.

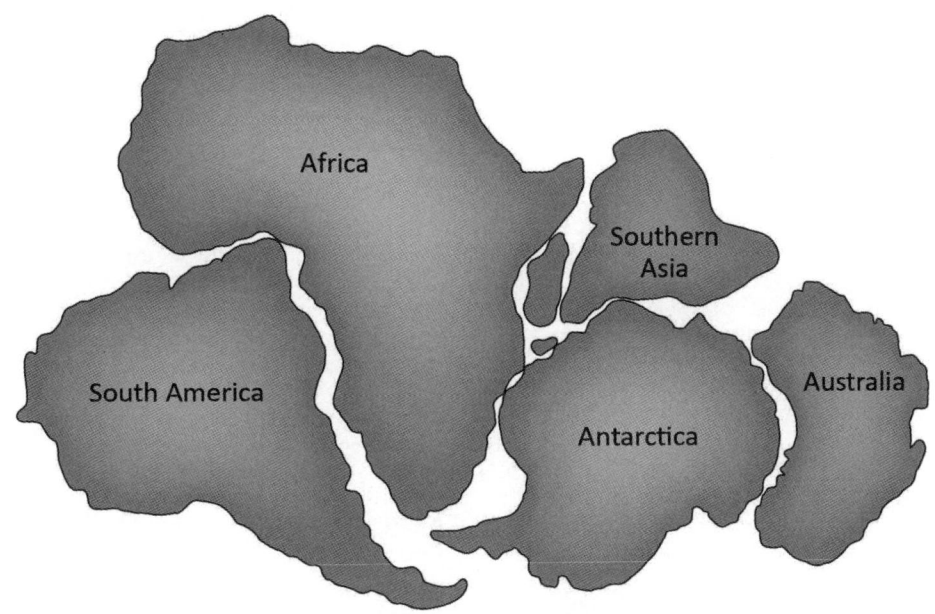

(b) Explain how fossils of *Cygnonathus* and *Glossopteris* can provide evidence for the scientists' theory.

....

....

[2]

(c) (i) Suggest why *Minmi* fossils **cannot** provide evidence for the theory.

....

....
[1]

(ii) Suggest why *Trilobite* fossils **cannot** provide evidence for the theory.

....

....
[1]

Periodic Table

1	2											3	4	5	6	7	0
						1 **H** hydrogen 1											4 **He** helium 2
7 **Li** lithium 3	9 **Be** beryllium 4											11 **B** boron 5	12 **C** carbon 6	14 **N** nitrogen 7	16 **O** oxygen 8	19 **F** fluorine 9	20 **Ne** neon 10
23 **Na** sodium 11	24 **Mg** magnesium 12											27 **Al** aluminium 13	28 **Si** silicon 14	31 **P** phosphorus 15	32 **S** sulfur 16	35.5 **Cl** chlorine 17	40 **Ar** argon 18
39 **K** potassium 19	40 **Ca** calcium 20	45 **Sc** scandium 21	48 **Ti** titanium 22	51 **V** vanadium 23	52 **Cr** chromium 24	55 **Mn** manganese 25	56 **Fe** iron 26	59 **Co** cobalt 27	59 **Ni** nickel 28	63.5 **Cu** copper 29	65 **Zn** zinc 30	70 **Ga** gallium 31	73 **Ge** germanium 32	75 **As** arsenic 33	79 **Se** selenium 34	80 **Br** bromine 35	84 **Kr** krypton 36
85 **Rb** rubidium 37	88 **Sr** strontium 38	89 **Y** yttrium 39	91 **Zr** zirconium 40	93 **Nb** niobium 41	96 **Mo** molybdenum 42	[98] **Tc** technetium 43	101 **Ru** ruthenium 44	103 **Rh** rhodium 45	106 **Pd** palladium 46	108 **Ag** silver 47	112 **Cd** cadmium 48	115 **In** indium 49	119 **Sn** tin 50	122 **Sb** antimony 51	128 **Te** tellurium 52	127 **I** iodine 53	131 **Xe** xenon 54
133 **Cs** caesium 55	137 **Ba** barium 56	139 **La*** lanthanum 57	178 **Hf** hafnium 72	181 **Ta** tantalum 73	184 **W** tungsten 74	186 **Re** rhenium 75	190 **Os** osmium 76	192 **Ir** iridium 77	195 **Pt** platinum 78	197 **Au** gold 79	201 **Hg** mercury 80	204 **Tl** thallium 81	207 **Pb** lead 82	209 **Bi** bismuth 83	**Po** polonium 84	**At** astatine 85	**Rn** radon 86
Fr francium 87	**Ra** radium 88	**Ac***** actinium 89	**Rf** rutherfordium 104	**Db** dubnium 105	**Sg** seaborgium 106	**Bh** bohrium 107	**Hs** hassium 108	**Mt** meitnerium 109	**Ds** darmstadtium 110	**Rg** roentgenium 111							

Key:
relative atomic mass
atomic symbol
name
atomic (proton) number

Elements with atomic numbers 112–116 have been reported but not fully authenticated

La lanthanoids

Ac actinoids

Glossary

absorb: to take in or soak up; to take in energy.
adaptation: characteristic of an organism that helps it to survive in a certain ecosystem.
aerobic respiration: respiration that requires oxygen to release energy from glucose.
alkali: a base that is soluble.
alkaline solution: a solution formed when a base dissolves in water.
amplitude: the maximum height of the wave, from the centre to the top or bottom.
anomalous: result that is very different from what you expect based on other results, perhaps because you made a mistake while recording it or something unexpected happened.
balanced forces: when the resultant force is zero.
basalt: a type of rock that contains quantities of minerals that can be magnetised.
base: a compound that can react with an acid and neutralise it.
bladder: organ that stores urine.
blood vessels: tube-shaped organs that carry blood around the body.
capture hypothesis: the idea that the Moon is a large asteroid that has been pulled into orbit around the Earth.
carbohydrate: nutrient needed for energy. Examples include starch and sugars (such as glucose).
chemical property: a property that is seen when a substance takes part in a chemical change.
chemical reaction: change in which new substances are produced.
chlorophyll: green substance that absorbs light, to get energy for photosynthesis.
chloroplast: green part of a cell that contains chlorophyll.
chromosome: structure containing a molecule of DNA, which carries genetic information in genes.
coastline: outside edge of a continent, where rock meets the ocean.

co-formation hypothesis: the idea that the Moon and Earth formed together, close to each other, at the same time.
collision hypothesis: the idea that a large object roughly the same size and mass as the planet Mars collided with the Earth, releasing rocks that were pulled together to form the Moon.
combustion: chemical reaction between a substance and oxygen, which transfers energy as heat and light.
compound: substance made from elements.
concentration: a measurement of how many particles of a certain type there are in a volume of liquid or gas.
conduction: form of heat transfer in which thermal energy passes through a substance from particle to particle. Conduction occurs mainly in solids.
conservation of energy: energy cannot be created or destroyed. The total amount of energy is constant.
constructive interference: this happens when two or more waves are added together to make a bigger wave.
continuous variation: variation that can have any value within a range.
convection: form of thermal transfer in which thermal energy causes a substance to expand and rise. This then cools and sinks. Convection only occurs in gases and liquids.
convection current: the movement of particles in a fluid due to convection.
covalent bond: a bond made when a pair of electrons is shared by two atoms.
crater: a roughly circular hole in the surface rock of a moon or planet caused by the impact of a large rock or asteroid.
crest: the highest point of a wave.
crystallisation: the formation of crystals from their solution.
Cynognathus: reptile that lived on Pangaea but became extinct over 175 million years ago; fossils found today provide evidence for plate movement.
data: numbers and words that can be organised to give information.
decomposition: process in which bacteria and fungi feed on dead animals and plants.

deduce: arrive at a logical conclusion based on available information.
density: the mass of an object divided by its volume.
destructive interference: this happens when two or more waves combine to make a smaller wave or to cancel each other out altogether.
diffusion: the spreading out of particles from where there are many (high concentration) to where there are fewer (lower concentration).
discontinuous variation: variation that has a distinct set of options or categories.
displacement reaction: chemical reaction in which a more reactive metal displaces (takes the place of) a less reactive metal from a compound, to form a new compound.
DNA: the substance that carries genetic information.
dot-and-cross diagram: a diagram used to show a covalent bond between atoms in a molecule. Electrons are represented by dots or crosses.
drought: unusually long period with low or no rainfall, causing water shortages.
drug: any substance that changes something about the way your body works.
ecosystem: all the organisms and the physical factors in an area.
egg cell: female gamete.
electrons: very small negatively charged particles in an atom.
element: substance that contains only one type of atom.
embryo: small ball of cells that develops from a fertilised egg cell. It becomes attached to the uterus lining and develops into a fetus.
emit: give out energy in the form of radiation.
endangered: a species is endangered if there are not many individuals left alive.
environment: the other organisms and physical factors around a certain organism.
environmental variation: variation in characteristics caused by an organism's surroundings.
evaporation: when a liquid turns into a gas, at a temperature lower than the boiling point. An evaporating liquid takes energy with it and so it cools the surface it was evaporating on; separating technique used to remove water from a solution.
evolution: a gradual change in something over time.
excretion: getting rid of wastes that are made inside an organism.
excretory system: organ system that removes wastes from the blood and produces urine.
extinction: when a species dies out completely.
filtration: separating technique used to remove an insoluble solid from a solution.
flood defences: walls, banks and other constructions designed to prevent flooding.
fluid: a substance that can flow from one place to another – a gas or a liquid.
formula: shows the chemical symbols of elements in a compound, and how many of each type of atom there are.
fossil fuels: remains of dead organisms that can be burned to release energy, including methane (natural gas), coal and oil.
fossils: remains of dead organisms from many millions of years ago that have solidified due to the pressure of sedimentary rocks forming above them.
frequency: the number of times an event occurs; the number of waves per second.
frequency diagram: any diagram showing the frequency of something.
gene: section of DNA that controls the development of a specific characteristic.
genetic material: substance found in a cell that controls how the cell develops and what it does. The genetic material of most organisms is DNA.
giant structure: an element or compound that is made up of atoms or ions joined together by strong bonds.
Glossopteris: plant that lived on Pangaea but became extinct over 175 million years ago; fossils found today provide evidence for plate movement.
glucose: sugar made by digesting carbohydrates (in animals) and by photosynthesis (in plants).
group: column in the Periodic Table.
guard cell: cell that helps form a stoma in a

leaf, to allow gases in and out.
hazard: harm that something may cause.
hertz: the unit of frequency. 1 Hz = one complete wave every second.
hydrocarbon: substance containing molecules made only of carbon and hydrogen.
inert: unable to take part in a chemical reaction.
infrasound: sound waves with a frequency too low for humans to hear.
inheritance: passing of features from parents to children.
inherited variation: variation in characteristics caused by an organism's parents.
insulator: a material which is a poor thermal conductor.
interference: what happens when two or more waves meet and their effects are added together.
invasive species: a non-native species that damages an ecosystem.
iodine solution: liquid that turns from orange to blue-black when added to starch.
ion: an atom which has gained at least one electron to be negatively charged or lost at least one electron to be positively charged.
ionic bond: an attraction between a positively charged ion and a negatively charged ion.
joule: the scientific unit for energy. Its abbreviation is J. 1000 J = 1 kilojoule (kJ); 1 000 000 J = 1 megajoule (MJ).
kidney: an organ that removes waste substances from the blood.
kinetic energy: energy stored by an object because it is moving.
Lystrosaurus: reptile that lived on Pangaea but became extinct over 175 million years ago; fossils found today provide evidence for plate movement.
magnetic alignment: occurs where rocks that are magnetised produce magnetic fields that line up with the Earth's magnetic field.
magnetic field reversal: process where the direction of Earth's magnetic field changes to be in the opposite direction.
mantle: deep layer of molten rock underneath the Earth's crust.

mass: the amount of matter in an object – it is measured in grams or kilograms.
mass extinction: when a very large number of species of living things become extinct over a short time.
matter: any substance that has mass, which is usually made up of atoms or molecules containing protons, neutrons and electrons.
Mesosaurus: reptile that lived in fresh water on and near Pangaea but became extinct over 175 million years ago; fossils found today provide evidence for plate movement.
meteor: an asteroid that enters the Earth's atmosphere.
meteorite: the piece of rock that is left behind after a meteor collides with the Earth's surface.
minerals: nutrients that living organisms need in small amounts for health, growth and repair. Also called mineral salts.
mitochondria: organelles (parts) in cells where respiration occurs.
molecule: a group of two or more atoms joined together.
natural selection: the process by which organisms have (by chance) better adaptations for new environmental conditions, making them more likely to survive and reproduce than other individuals of that species.
nebula (plural nebulae): a cloud of interstellar gas and dust.
neutrons: particles with no charge in the nucleus of an atom.
nuclear fusion: process in which the nuclei of two atoms are merged together, releasing large amounts of energy.
nucleus: the central part of an atom – contains protons and neutrons.
nutrition: life process by which plants and animals take in and break down substances and use them to get the nutrients needed for other life processes.
organ system: group of organs working together.
Pangaea: supercontinent on Earth that broke apart about 175 million years ago.
parallel circuit: a circuit made up of more than one loop.

period: row in the Periodic Table.
Periodic Table: how the elements are arranged, in order of their atomic number.
photosynthesis: a series of chemical reactions in which carbon dioxide and water are converted to glucose and oxygen.
physical property: the property that can be observed or measured without changing the basic nature of the substance.
pitch: how high or low a sound is.
pollutant: a substance in an ecosystem that can cause harm to organisms.
pollution: when a substance in an ecosystem causes harm to organisms.
population: the total number of individual organisms of one species living in a certain area.
potential energy: the amount of stored energy an object has because of its position.
predator: animal that hunts and eats other animals (called prey).
prey: animal that is hunted and eaten by other animals (called predators).
product: substance made during a chemical reaction.
protons: positively charged particles in the nucleus of an atom.
radiation: form of energy transfer in which thermal energy is released as infra-red radiation. There is no change in matter for energy to transfer in this way.
range: the highest and lowest values in a set of data.
rate of reaction: how fast a chemical reaction happens.
raw material: another term for reactant.
reactant: substance that changes in a chemical reaction to form products.
reactivity: how likely it is that a substance will undergo a chemical reaction.
reactivity series: series of metals written in order from the most reactive to the least reactive.
red supergiant: a huge, red-coloured star that is formed when a massive star expands towards the end of its life.
reliable: measurements are reliable when repeated measurements give results that are very similar.
renal system: another name for the excretory system.
repeatable: results that are the same each time they are taken, when the same method and equipment are used.
resistance: a measure of how difficult it is for current to flow. Measured in ohms.
resistor: a device which resists the flow of current.
respiration: process by which organisms release energy through the conversion of glucose and oxygen to carbon dioxide and water.
resultant force: shows the single total force acting on an object when all the forces acting on it are added up.
risk: chance of a hazard causing harm.
robotic: describes a device that works by itself, without a human needing to control its movements.
root hair cell: plant cell found in roots that is adapted for taking in water quickly.
salt: a type of compound that consists of metal atoms joined to non-metal atoms, e.g. sodium chloride.
series circuit: a circuit made up of a single loop.
shells: the paths or orbits that electrons move along in an atom.
shock wave: the squeezing of a volume of air caused by an explosion, resulting in a wave of high pressure that spreads out from the explosion.
simple structure: an element or compound that is made up of molecules.
sink: something that takes in and stores significant quantities of a particular substance (for example, plants act as a sink for carbon dioxide).
source: something that produces significant quantities of a particular substance (for example, animals and plants act as a source for carbon dioxide).
starch: large, insoluble carbohydrate made by plants to store energy and an important energy source in human diets.
stellar nursery: the dense part of a nebula in which new stars are formed.
stillborn: the term used to describe a baby that is dead when it is born.
stoma (plural stomata): hole in a leaf, formed between two guard cells.

sugar: soluble carbohydrate, which exists as small molecules. Glucose is an example.
supercontinent: area of land where a number of continents were joined together.
supernova: the explosion of a massive star at the end of its life, which produces a nebula.
surface area: the area of a surface, measured in squared units such as square centimetres (cm^2).
symbol equation: way of showing a chemical reaction using formulae – a balanced symbol equation has equal numbers of each type of atom on both sides of the equation.
tally chart: a table used to help count things.
temperature: a measure of how hot or cold something is. It is the average amount of thermal energy in a substance.
theory: idea or set of ideas that explains an observation.
theory: a hypothesis with enough supporting evidence that it is thought to be the most likely explanation for a process.
thermal energy: energy stored in an object due to its temperature.
transpiration: the loss of water vapour through the stomata on the surface of the leaves.
trough: the lowest point of a wave.
ultrasound: sound waves with a frequency too high for humans to hear.
unbalanced forces: when there is a resultant force.

upthrust: the upwards force on an object from the liquid or gas in which it is floating.
urea: main waste substance removed from the blood by the kidneys.
ureter: tube-shaped organ that carries urine from a kidney to the bladder.
urethra: tube-shaped organ that carries urine from the bladder to outside of the body.
urinate: to release urine from the bladder.
urine: liquid containing urea and other wastes.
variation: differences between characteristics.
voltage: a easure of energy in a circuit.
voltmeter: device for measuring voltage.
volume: the amount of space an object takes up, measured in cm^3.
waveform: the shape of a wave.
wavelength: the length of one complete wave.
whole number: number without fractions or a decimal point.
wild fires: large fires in natural areas, which spread quickly due to dry and dead plants, often made worse in a drought.
wilting: when a plant becomes floppy due to lack of water.
word equation: model showing what happens in a chemical reaction, with reactants on the left of an arrow and products on the right.
xylem cell: plant cell that is adapted to form hollow tubes to transport water.
xylem vessel: tube formed by the joining of many dead xylem cells.